Study Guide and Workbook

for
Masterton and Hurley's

Chemistry
Principles and Reactions

Updated Fifth Edition

Study Guide and Workbook

for
Masterton and Hurley's

Chemistry
Principles and Reactions

Updated Fifth Edition

Cecile N. Hurley
University of Connecticut

THOMSON

BROOKS/COLE

Australia • Canada • Mexico • Singapore • Spain • United Kingdom • United States

Printer: Quebecor World/Dubuque

0-495-01141-X

For more information about our products,
contact us at:
Thomson Learning Academic Resource Center
1-800-423-0563

For permission to use material from this text or product, submit a request online at
http://www.thomsonrights.com.
Any additional questions about permissions can be submitted by email to **thomsonrights@thomson.com.**

Thomson Higher Education
10 Davis Drive
Belmont, CA 94002-3098
USA

Asia (including India)
Thomson Learning
5 Shenton Way
#01-01 UIC Building
Singapore 068808

Australia/New Zealand
Thomson Learning Australia
102 Dodds Street
Southbank, Victoria 3006
Australia

Canada
Thomson Nelson
1120 Birchmount Road
Toronto, Ontario M1K 5G4
Canada

UK/Europe/Middle East/Africa
Thomson Learning
High Holborn House
50–51 Bedford Road
London WC1R 4LR
United Kingdom

Latin America
Thomson Learning
Seneca, 53
Colonia Polanco
11560 Mexico
D.F. Mexico

Spain (including Portugal)
Thomson Paraninfo
Calle Magallanes, 25
28015 Madrid, Spain

Preface

Collaborative (or, cooperative) learning is not new. Most general chemistry students bring ample experience with it from their elementary and high-school careers. What *is* new is its growing use in colleges and universities as a means of promoting improved learning. Cooperative learning does that in several ways, including

- providing a new, more active structure for weekly discussion sections

- giving students the opportunity to verbalize their thought processes clearly enough to make them understandable to their group peers

- exposing students to new and different ways that their peers may use to solve problems, visualize concepts, and think about how to apply theory

- encouraging students to work together on problems that promote deeper understanding than routine homework questions.

Studies at the University of Connecticut show that students in sections with collaborative learning have averaged half a letter grade higher than those in standard sections of the same course. This approach has been particularly effective for women, spectacularly so for minority women. Consequently, all sections of our general chemistry course now use group learning in the weekly discussion period. The results of our study (which presents a detailed statistical analysis of the experinece with collaborative learning at the University of Connecticut) can be found in the article "Cooperative Learning in the Quiz Section in General Chemistry" which appears in *Proc. Frontiers in Education 23,* Washington, DC, IEEE & ASEE, 1993, 162-166.

The focus of our cooperative learning is the weekly worksheet. This manual has three different ones for Chapters 1–19 of the text. (Chapters 20–22 are largely descriptive and use the principles discussed in earlier chapters. There are no worksheets for these three chapters.) The section *Organization* discusses various approaches to breaking a class up into such groups. Teams of 3-to-5 students discuss the worksheet each week under the supervision of an instructor or teaching assistant.

If students get "stuck" or ask for clarification of a point, the instructor provides a hint or asks leading questions to get them moving again. It is *very* rare for them to coach students actively on how to analyze a problem, and rarer still for them to present a step-by-step solution. Instead, when the groups complete an item, they volunteer an answer and ask the instructor to appraise the reasoning that led them to it.

The worksheets are all of comparable difficulty. Each one is roughly half conceptual and half quantitative, with the latter part more challenging than the assigned homework problems the students work out before they come to discussion. It should take the student teams roughly an hour to do the entire worksheet. Our experience has led us to divide the worksheets into 3 or 4 parts, assigning a part to each group of 3 to 5 students. The groups work on their assignignments for about 40 minutes and spend the rest of the hour presenting their solutions and answers to ther assigned question to the rest of the class.

The conceptual half of each worksheet aims to teach students to be precise in both their thinking and their language. They learn that words such as *never* and *always* are rarely valid in chemistry, and that sweeping generalizations are risky. They teach each other that decreasing *by* 25% is not the same as decreasing *to* 25%. It is in this area of chemical reasoning that we have seen the students make the greatest improvement. The class mean on the conceptual part of the hour exams used to hover in the range of 50-to-60%. After we switched to group learning during the quiz sections, that climbed dramatically — to a range of 68-to-75%.

These worksheets have been used in other ways by instructors at other universities. Some use it as the framework for a quiz section where the whole class participates (as a group) in discussing the entire worksheet with the TA (or instructor) as group leader. If the format of your quiz section is such that a TA is responsible for the activity that takes place in his/her discussion section, then these worksheets are particularly useful to the new TA's who have no experience in selecting the type of questions that should be discussed.

Other instructors assign the worksheet as a group take-home exam. Still others assign them for extra-credit work or as a make-up for missed assignments. However you choose to use them, it is our experience that a certain amount of time should be spent listening to the student's reasoning on how he/she arrived at the answers handed in. These worksheets were never intended to substitute for written homework problems. Their efficacy seems to be largely due to having the students verbalize their reasoning process.

It is a pleasure to acknowledge the contributions of individuals and institutions to this endeavor. The development of the collaborative approach to general chemistry, which stimulated the creation of these worksheets, was supported in part by a grant from the National Science Foundation Curriculum Development Program in Chemistry (USE – 9155980). That grant also provided partial support for the testing and assessment of the efficacy of the new instructional mode. Joel Tolentino and Maria Cecilia de Mesa reviewed these worksheets for accuracy. James F. Hurley served as sounding board, critic, and booster. While the deficiencies are my sole responsibility, the production of these worksheets in final form owes much to their efforts.

Cooperative learning is an exciting tool that has shown it can be of real value in general chemistry. It can help instructors as well as students, and I hope that these worksheets can help you implement a successful program of group work in your class.

– Cecile N. Hurley
Storrs, Connecticut
November, 2004

To the memory of
Gary A. Epling & Edward L. Kostiner

Thank you for letting me stand on your shoulders

How to Use This Workbook

What Is It?

This workbook is designed to help you learn general chemistry. It should be used in conjunction with *Chemistry: Principles & Reactions* by Masterton and Hurley. It is *not* a substitute for either your professor's lectures or the text.

How Is It Organized?

Each chapter outlines the material in the corresponding chapter of the text. The outlines contain fill-in blanks, worked examples, exercises, and various warnings, mnemonics, and suggestions for analyzing and solving problems. You will find a self-test at the end of most chapters. All pages are scored, so they can easily be torn out. They also have prepunched holes to fit in a standard three-ring binder.

What should you do with it?

To get the most from this workbook, you should attend all your professor's lectures and try to take complete notes. As soon as possible after each lecture — preferably the same day — study your notes. Find the part of the chapter outline in this workbook that corresponds to those notes. Tear it out, and read through it. If you can, fill in the blanks from memory. Study the examples, and then try to work all the exercises. If you cannot do some of the exercises, *seek help at once from your professor or TA.* **Do not** wait until the end of the chapter. Insert all the torn-out sheets you have used into your binder directly after your class notes. You will then be ready to try the assigned problems from the text.

An effective way to deal with the assigned homework problems and to master the unassigned problems is to form a study group with a few of your classmates. Make each member of the study group responsible for certain designated problems each week, and hold a weekly meeting of your study group to go over all the problems in detail. That way, you and your fellow group members can get correct answers and solution methods for all the problems you aim to cover. Be sure to schedule a regular meeting at the same time each week, so that you will get into the study-group habit. This approach to studying general chemistry has been very helpful for many students at the University of Connecticut.

To review for an exam, restudy your notes and tear-out sheets for each chapter that the exam will cover. After studying each chapter, take its self-test. Do that under exam conditions: a copy of the Periodic Table and a calculator should be all you need. Do *not* refer to your notes or to the text, except to get information from a table.

If you follow these suggestions, you should find general chemistry easier to understand and master, and more enjoyable.

Contents

1

Matter and Measurements

I. Types of Matter
  A. Elements
    Definition: _____
  B. Compounds
    1. Definition: _____

      _____

    2. Resolving compounds into compound elements
      a. Heat
      b. Electrolysis
  C. Mixtures
    1. Definition: _____

      _____

    2. Types
      a. Homogeneous (definition): _____

        _____

      b. Heterogeneous (definition): _____

        _____

    3. Separation techniques
      a. Filtration
        (1) Kind of mixture separated: _____

        (2) Brief description: _____

          _____

      b. Distillation
        (1) Kind of material separated: _____

(2) Brief description: _____

_____

   c. Gas-liquid chromatography

     (1) What difference in the components of the mixture does it take advantage of? _____

_____

     (2) Brief description: _____

_____

     (3) Application: _____

II. Measurements

  A. Metric System

Prefix	Factor	Abbreviation
kilo	_____	_____
centi	_____	_____
milli	_____	_____
micro	_____	_____
nano	_____	_____
pico	_____	_____

  B. Instruments and Units

    1. Length

      a. Instrument: _____

      b. Standard unit in the metric system: _____

    2. Volume

      a. Three common units used:

        (1) Cubic centimeter

        Abbreviation: _____

        Relation to 1 liter _____

        (2) Liter

        Abbreviation: _____

        Relation to 1 m^3 _____

      (3) Milliliter

          Abbreviation: _____

          Relation to 1 L _____

   b. Device most commonly used to measure volume: _____

3. Mass

   a. Units used to express mass in the metric system are:

    _____ , _____ , _____ .

    A metric ton = _____ g.

   b. Difference between mass and weight

    Mass: _____

    _____

    Weight: _____

    _____

4. Temperature

   a. Unit

      (1) Celsius – °C

      (2) Fahrenheit – °F

      (3) Kelvin – K

   b. Relationships of different units to each other

      (1) Celsius and Fahrenheit

      (2) Celsius and Kelvin

   c. Instrument for measurement: _____

   d. *Exercises*

     (1) During the lunar month, when the surface of the moon is heated by the sun, the temperature of the moon reaches a high of 251°F. Express that temperature in °C and K.

     Solution: The relationship between °F and °C is expressed by the equation

$$°F = 1.8(°C) + 32$$

Substituting we get

$$251 = 1.8(°C) + 32$$

Subtracting 32 from both sides and dividing both sides by 1.8, we obtain

$$°C = \frac{251 - 32}{1.8} = 122$$

The relationship between °C and K is

$$K = °C + 273$$

Substituting and adding, we obtain

$$K = 122 + 273 = 395$$

(2) During the dark period of the lunar month the temperature drops to a low of -137°C. Express that temperature in °F and K. (**E1**)

III. Significant Figures
A. Definition: _____
B. Rules governing the manipulation of significant figures:
   1. Counting the number of significant figures:
      a. 2.698 has four significant figures.
      RULE: _____

      _____

      b. There are six significant figures in 100374 mm as well as in 10.0204 mm.
      RULE: _____

      _____

      c. There are four significant figures in 1.800 L and three significant figures in 1.80 L.
      RULE: _____

      _____

      d. There are two significant figures in 0.00039 J and three significant figures in 0.0398 J.
      RULE: _____

      _____

   2. Multiplication and division:
      RULE: _____

      _____

   3. Addition and subtraction:
      RULE: _____

      _____

4. Significant figures and exact numbers:
  RULE: _____

  _____

5. *Exercises*

   a. How many significant figures are there in

   (1) 0.008090 mL?
   Solution: There are four significant figures. The zero before the decimal point and the two zeros after the decimal point are not significant; 8, 0, 9, and 0 are all significant. To check: 0.008090 in exponential notation is 8.090×10^{-3}. This clearly has four significant figures.

   (2) 1300.40 atm? **(E2)**

   (3) 0.0010960 m? **(E3)**

   (4) one liter = 1000 mL? **(E4)**

   b. Assuming the numbers given represent measured quantities, calculate

   (1) $\dfrac{3.016 \times 4.23}{0.0031}$

   Solution: The measurement with the least number of significant figures is 0.0031 (two significant figures). Hence, the answer should have two significant figures. If you work out the calculations on your calculator, you will see 4115.3806. The answer with the correct number of significant figures is then 4.1×10^3. It is not a good idea to write 4100 since someone reading your answer might conclude that you mean to have four significant figures.

   (2) $\dfrac{8.13 \times 0.006976}{3.9874}$ **(E5)**

c. Calculate the total mass of the following items:

Beaker	37.012 g
Textbook	2173.1 g
Weighing paper	0.00389 g

Solution: The calculator readout shows 2210.11589. According to the rules of addition and subtraction, the uncertainty of the result is determined by the absolute uncertainty of the least precise measurement. In this case that measurement is the mass of the textbook, which is accurate only to the tenth of a gram. The answer, therefore, should be accurate only to a tenth of a gram, and thus is 2210.1 g.

d. A mixture of sulfur and iron has a mass of 13.6841 g. A magnet separates the iron, whose mass is 0.286 g. What is the mass of the sulfur? **(E6)**

e. An experiment calls for 12.28 g of sodium chloride. The instructions are to double that amount. How many grams of sodium chloride do you need? **(E7)**

IV. Unit Conversion

A. Steps in doing unit conversions:

1. Choose a factor that cancels the unit you want to eliminate.

2. Choose a factor that has the unit you want to obtain.

(Many times you can do Steps 1 and 2 with one conversion factor. At other times it may require more than one factor.)

3. Multiply, making sure your units cancel.

B. *Exercises*

1. Convert 3.0 nm to meters.

Solution: Conversion factor: 1 nm (unit to be eliminated) = 10^{-9}m (unit you want).

$$3.0\,\text{nm} \times \frac{10^{-9}\text{m}}{1\,\text{nm}} = 3.0 \times 10^{-9}\text{m}$$

Note that the answer has two significant figures because 3.0 nm has two significant figures and the conversion factor 1 nm = 10^{-9}m is exact.

2. Convert 36.4 feet to miles. **(E8)**

3. Convert 0.416 miles to centimeters.
   Solution: Although we do not know a conversion factor that directly changes miles to centimeters, we do know the following conversion factors (hopefully without having to look them up!):

$$1 \text{ mile} = 5280 \text{ feet}$$
$$12 \text{ inches} = 1 \text{ foot}$$
$$2.54 \text{ centimeters} = 1 \text{ inch}$$

Thus, the first thing to do is to eliminate the unit miles. We use the conversion factor that involves miles; that is, 1 mile (mi) = 5280 ft.

$$0.416 \, \text{mi} \times \frac{5280 \, \text{ft}}{1 \, \text{mi}}$$

We have eliminated miles but have feet as units, which we also do not want. So, we use another conversion factor to remove feet.

$$(0.416 \times 5280) \, \text{ft} \times \frac{12 \, \text{in}}{1 \, \text{ft}}$$

Now we have units in inches, which is still not centimeters, but the next conversion factor will convert directly to centimeters.

$$(0.416 \times 5280 \times 12) \, \text{in} \times \frac{2.54 \, \text{cm}}{1 \, \text{in}}$$

Our whole equation will thus look like this:

$$0.416 \, \text{mi} \times \frac{5280 \, \text{ft}}{1 \, \text{mi}} \times \frac{12 \, \text{in}}{1 \, \text{ft}} \times \frac{2.54 \, \text{cm}}{1 \, \text{in}} = 66948.7104$$

Your calculator readout will probably show 66948.7104. The answer with the correct number of significant figures is 6.69×10^4 cm. This method may seem cumbersome at first, but if you use it exclusively from now on, you will find that it relieves you of wondering, as you did in high school, whether you should multiply or divide. It will also make unnecessary the use of ratios and proportions that may cause trouble later on in the course.

4. Convert 3897.2 lb into metric tons. **(E9)**

5. A certain drug is metabolized at a constant rate of 0.0236 oz/hr. Convert this to mg/min. **(E10)**

V. Properties
   A. Identifying properties
      1. Intensive
         a. Definition: _____
         b. Example: _____
      2. Extensive
         a. Definition: _____
         b. Example: _____
   B. Chemical properties
      1. Definition: _____
      2. Example: _____
   C. Physical properties
      1. Definition: _____
      2. Density
         a. Definition: _____
         b. Units: _____
         c. Mathematical equation:

$$\text{Density} = \frac{\text{mass}}{\text{volume}}$$

d. Exercises

(1) A student wishes to determine the density of a solution of sodium hydroxide. He finds the mass of the flask to be 25.631 g. When he fills the flask with water at 22°C (d = 1.00 g/cm³), its mass is 35.432 g. He finds that the mass of the flask filled with sodium hydroxide solution is 36.138 g. What is the density of the sodium hydroxide solution?

Solution: In this problem, we can easily determine the mass of the solution. It is (mass of flask + solution) − (mass of empty flask). Thus

$$\text{mass of solution} = 36.138\,\text{g} - 25.631\,\text{g}$$
$$= 10.507\,\text{g}$$

The volume of the solution is the volume of the flask. It is also the volume of the water used to fill the flask. The mass of the water filling the flask can be determined as above where the mass of the water is equal to the mass of the flask plus water minus the mass of the flask. Thus

$$\text{mass of water} = 35.432\,\text{g} - 25.631\,\text{g}$$
$$= 9.801\,\text{g}$$

The volume of the water is 9.80 cm³ since the density of water is given as 1.00 g/cm³. This is also the volume of the flask and the volume of the sodium hydroxide solution. The density of the solution is therefore

$$d = \frac{\text{mass}}{\text{volume}} = \frac{10.507\,\text{g}}{9.80\,\text{cm}^3} = 1.07\ \text{g/cm}^3$$

(2) What is the mass of a flask filled with acetone (d = 0.792 g/cm³) if the same flask filled with water (d = 1.00 g/cm³) weighs 75.2 g? The empty flask weighs 49.74 g. **(E11)**

(3) Ten grams of zinc pellet ($d = 7.14$ g/cm^3) are added to a flask (123.4 g) filled with water. The flask and water ($d = 1.00$ g/cm^3) weigh 211.6 g. How much does the flask weigh after the zinc is added and the sides dried from the displaced water? **(E12)**

3. Melting point
  a. Definition: _____

  b. Unit: _____
4. Boiling point
  a. Definition: _____

  b. Unit: _____
5. Solubility
  a. Definition: _____
  b. Unit: Most of the time it is grams of compound/100 g of water. This unit can be made into a conversion factor and expressed as

$$\frac{\text{grams of compound}}{100 \text{ g of water}} \quad \text{or} \quad \frac{100 \text{ g of water}}{\text{grams of compound}}$$

  c. Solubility is affected by temperature. The solubility of some compounds increases with an increase in temperature. Other compounds have a decrease in solubility with an increase in temperature.

  d. *Exercises*
  (1) The solubility of potassium nitrate at 70°C is 137 g/100 g water. How much water would be needed to dissolve 17.58 g of potassium nitrate at 70°C?
  Solution: Here we use the unit conversion method. Our conversion factor is

$$\frac{137 \text{ g potassium nitrate}}{100 \text{ g water}}$$

We want the answer to be expressed in grams of water. Hence

$$17.58 \text{ g potassium nitrate} \times \frac{100 \text{ g water}}{137 \text{ g potassium nitrate}} = 12.8 \text{ g H}_2\text{O}$$

(2) It is determined that 22 g of sodium chloride are soluble in 60.0 g of water at 40°C. What is the solubility of sodium chloride in grams sodium chloride/100 g water? **(E13)**

(3) The solubility of potassium nitrate (KNO_3) is 38 g/100 g H_2O at 20°C and 1.50×10^2 g/100 g H_2O at 70°C. A solution is prepared by mixing 45 g of KNO_3 in 35 g of water at 70°C.

(a) Will all the potassium nitrate dissolve at 70°C?

(b) If it does, how much more solute can be dissolved at that temperature? If it does not, how much solute did not dissolve?

(c) The solution is cooled to 20°C. Does all the KNO_3 remain in solution at this temperature?

(d) If it does, how much more KNO_3 can be dissolved at 20°C? If it does not, how much KNO_3 crystallizes out? **(E14)**

6. Color
   a. The colors of gases and liquids are due to _____ .

   _____

   b. When violet color is absorbed, the color transmitted is _____ .

   c. The range of wavelengths for visible light is _____ .

SELF-TEST
A. Multiple choice:

1. Which one of the following could be described as a chemical property?
   a. The crystal is rhombic in shape.
   b. The density is 2.593 g/cm^3.
   c. The solubility is 43 g/100 g H_2O at 45°C.
   d. The compound reacts violently with water.

2. A compound X has a solubility of 79 g/100 g H_2O. When 24 g of X are added to 68 g of water, the resulting mixture is
   a. a homogeneous solution.
   b. a heterogeneous mixture.
   c. a compound.
   d. an element.

3. Chicken noodle soup is
   a. a heterogeneous mixture.
   b. an element.
   c. a homogeneous solution.
   d. a pure compound.

4. The grocer sells you 2 dozen eggs. He tells you that he will give you 25 eggs for the price of 2 dozen. The number of significant figures in 25 is
   a. 1   b. 2   c. ambiguous   d. infinite

5. The solubility of compound Y is plotted against temperature. A straight line with a negative slope is obtained. This means that
   a. the solubility of Y increases with increasing temperature.
   b. the solubility of Y decreases with increasing temperature.
   c. the solubility of Y is independent of temperature.
   d. all of the above are true.

B. Answer the questions below, using **LT** (for *is less than*), **GT** (for *is greater than*), **EQ** (for *is equal to*), or **MI** (for *more information required*) in the blanks provided.

_____ 1. Consider two compounds A and B of equal mass. The density of A is larger than the density of B. The volume of A __(1)__ the volume of B.

_____ 2. The number of significant figures for x in the following calculation is __(2)__ 2.

$$\frac{(2.68)(1.9) - (0.4)(0.01396)}{0.7143}$$

_____ 3. A compound Z has a solubility of 39 g/100 g H_2O at 45°C. Its solubiity at 90°C is __(3)__ 39 g/100 g H_2O.

_____ 4. The temperature of lukewarm water is about 52°C. If a thermometer calibrated to a tenth of a degree is used to measure the temperature of the water, the number of significant figures in the measurement is __(4)__ 3.

_____ 5. A cork pellet is dropped into water and floats. The density of the cork pellet is __(5)__ 1.

_____ 6. 0.00123 mg __(6)__ 1.23 g.

C. Problems:
Consider bromine with the following properties:
density: 3.12 g/cm³
melting point: −7°C
boiling point: 59°C
solubility: 3.3 g/100 g water at 25°C, 1 atm
color: red

1. How many pounds of bromine would fill 1.68 qt?

2. A student tries to determine the density of a pebble by bromine displacement. She obtained the following data:

mass of flask filled with bromine	76.83 g
mass of flask filled with bromine + pebble	89.68 g
mass of pebble	23.45 g

What is the density of the pebble?

3. At 25°C, how many milliliters of bromine could you dissolve in 1.000 L of water? (Assume the density of water is 1.00 g/cm³.)

4. If we use the melting and boiling points of bromine for a new temperature unit, say, °B, where the melting point of bromine is 0°B and its boiling point is 100°B, derive the mathematical formula for converting °B to °C.

ANSWERS
Exercises:

(E1) −215°F, 136 K　　**(E2)** 6　　　　　　**(E3)** 5　　　　　　**(E4)** infinite
(E5) 0.0142　　　　　　**(E6)** 13.398 g　　**(E7)** 24.56 g　　**(E8)** 0.00689 miles
(E9) 1.768 metric tons　**(E10)** 11.2 mg/min　**(E11)** 69.9 g　**(E12)** 220.2 g
(E13) 37 g sodium chloride/100 g water　　**(E14)** yes; 7 g more; no; 32 g

Self-test
A. Multiple Choice:
　1. d　　　　　**2.** a　　　　　**3.** a　　　　　**4.** d　　　　　**5.** b

B. LT, GT, EQ, or MI
　1. LT　　　**2.** EQ　　　**3.** MI　　　**4.** GT　　　**5.** LT　　　**6.** LT

C. Problems
　1. 10.9 lb　　**2.** 6.92 g/mL　　**3.** 11 mL　　**4.** °C = 0.66°B − 7

2

Atoms, Molecules, and Ions

I. Atomic Theory
  A. Postulates
   1. _____

   _____

   2. _____

   _____

   3. _____

   _____

  B. Basic laws of chemistry explained by postulates of Atomic Theory
   1. Law of conservation of mass

   _____

   _____

   2. Law of constant composition

   _____

   _____

   3. Law of multiple proportions

   _____

   _____

II. Components of the atom
  A. Electron
   1. Location: _____

   2. Relative charge: _____

3. Relative mass: _____

B. Nucleus
   1. Discovery
      a. Scientist: _____
      b. Experiment: _____

   _____

   _____

   2. Components
      a. Proton
         (1) Relative charge: _____
         (2) Relative mass: _____
      b. Neutron
         (1) Relative charge: _____
         (2) Relative mass: _____

C. Atomic number
   1. Definition: _____

   _____

   2. Symbol: _____
   3. Relation of atomic number to electrons in a neutral atom:

   _____

D. Mass number
   1. Definition: _____

   2. Symbol: _____
   3. Isotopes
      Definition: _____
   4. Nuclear Symbol
      a. ^A_Z Symbol for element
      b. *Exercises*
         (1) Write the nuclear symbol for three isotopes of silicon ($Z = 14$) in which
            there are 14, 15, and 16 neutrons, respectively.
            Solution: For the isotope with 14 neutrons,

$$A = \text{no. of protons} + \text{no. of neutrons}$$
$$= 14 + 14 = 28$$

The nuclear symbol is:

$$\begin{matrix} A \to \\ Z \to \end{matrix} \quad ^{28}_{14}\text{Si}$$

For the isotope with 15 neutrons, the nuclear symbol is

$$^{29}_{14}\text{Si}$$

For the isotope with 16 neutrons, the nuclear symbol is

$$^{30}_{14}Si$$

(2) One of the isotopes of aluminum (Al) has the nuclear symbol $^{27}_{13}Al$.
   (a) How many protons are there in the nucleus?
   (b) How many neutrons?
   (c) How many electrons are there in the atom? **(E1)**

E. Nuclear Stability
   1. Stability of the Nucleus
      a. It is dependent on _____ ratio.
      b. Light elements can be predicted to have stable nuclei when _____
      _____ .
      c. Heavy elements can be predicted to have stable nuclei when _____
      _____ .
      d. Unstable isotopes decay by a process referred to as _____ .
   2. Forms of radiation
      a. β - particles
         These are particles with properties identical to those of _____ .
      b. α - particles
         These are _____ nuclei carrying a _____ charge
         because _____ .
      c. gamma rays
         These consist of _____ .

III. The Periodic Table
   A. The chemical properties of elements depend upon their atomic number (Z), which can be read off the periodic table (inside front cover of your textbook). Z is given directly above the symbol of the element.
   B. Organization
      1. Period (Definition): _____
      _____

      2. Group
         a. Definition: _____
         _____
         b. Designation
            (1) Groups are numbered from 1 to **18** starting at the left.
            (2) Group numbers of main-group elements: _____
            (3) Group numbers of transition elements: _____

   c. Groups with special names
     (1) Alkali metals
       (a) Group number: _____

       (b) Example: _____

       (c) Chemical property: _____

       _____

     (2) Alkaline earth metals
       (a) Group number: _____

       (b) Example: _____
     (3) Halogens
       (a) Group number: _____

       (b) Example: _____
     (4) Noble gases
       (a) Group number: _____

       (b) Example: _____

       (c) Chemical property: _____

       _____

3. Metals, nonmetals, and metalloids
   a. Diagonal line or stairway that starts to the left of boron (B) separates the metals from the nonmetals.
   b. Location of metals relative to the stairway:

     _____

   c. Location of nonmetals relative to the stairway:

     _____

   d. Metalloids
     (1) Location relative to the stairway:

       _____

     (2) What is a metalloid?

       _____

     (3) Name the six elements classified as metalloids.

       _____

IV. Molecules
  A. Nature
   1. Charge: _____

   2. Type of atoms that make up a molecule: _____

   3. Forces that hold atoms together are called _____

     and consist of _____ .

B. Representation
1. Structural formulas

    Definition: _____

2. Condensed structural formulas

    a. Definition: _____
    b. Reactive group

        _____ is the reactive group in alcohols.

        _____ is the reactive group in amines.

3. Molecular formulas
    a. Definition: _____
    b. *Exercises*
        (1) Write the molecular formula for the molecule made up of two atoms of
              hydrogen (H) and two atoms of oxygen (O).
              Solution:

$$H_2O_2$$

The subscripts after the element's symbol show how many atoms of
that element are in the molecule.
        (2) Write the molecular formula for
            (a) the molecule made up of one carbon (C) atom and four chlorine
                  (Cl) atoms. **(E2)**

            (b) the molecule made up of two carbon (C) atoms and four hydrogen
                  (H) atoms and two oxygen (O) atoms. **(E3)**

C. Behavior of molecular solutions
    An aqueous solution with a molecule as solute cannot conduct electricity. This type
    of solute is known as a(n) _____ .
V. Ions
    A. Definition: _____
    B. Types of ions
        1. Cations (definition): _____

            _____

        2. Anions (definition): _____

            _____

3. Number of protons and electrons in an ion
   a. The number of protons (atomic number) remains unchanged (as compared to the atom).
   b. The number of electrons changes
      (1) For cations (positive charge), the number of electrons decreases by the size of the charge.
      (2) For anions (negative charge), the number of electrons increases by the size of the charge.

      A good mnemonic for remembering this is *a*nions *i*ncrease the number of their electrons (the vowels go together) and *c*ations *d*ecrease the number of their electrons (the consonants go together).
   c. *Exercises*
      (1) Give the symbol for the ion that has 11 protons and 10 electrons.
            Solution: The atomic number (no. of protons) is 11, hence, the element is sodium (Na). The number of electrons decreases, therefore, the ion is a cation (positive charge) and the number of electrons is decreased by 1. Thus the charge must be +1. The symbol is therefore Na^+.
      (2) How many protons and electrons are there in the Al^{3+} ion? **(E4)**

      (3) How many protons and electrons are there in the S^{2-} ion? **(E5)**

4. Dissolving ionic solids in water
   a. When ionic solids dissolve in water, the solution contains _____ .
   b. The presence of charged particles allows the solution to conduct electricity.
   c. This ability to conduct electricity makes the solution a(n) _____ .

VI. Formulas for Ionic Compounds
   A. Nature of ionic compounds
   1. They are made up of two oppositely charged ions.
   2. The principle of electrical neutrality when applied to ionic compounds requires

      that _____

      _____ .

   3. The formula for an ionic compound shows the simplest ratio between cation and anion.

B. Ions with a noble-gas structure

   1. For main-group elements, the charges of the ions formed by the atoms can be predicted by applying this principle:

   _____

   _____

   2. Charge of ions formed by the main-group elements

   a. Group 1: _____

   b. Group 2: _____

   c. Group 16: _____

   d. Group 17: _____

   e. Al in Group 13: _____

   f. N in Group 15: _____

C. Cations of the transition and post-transition metals

   1. Transition metals are in groups 3 – 12.

   2. Post-transition metals are those in periods 4 – 7 and in groups **13, 14,** and **15.**

   3. These cations typically have charges of _____ .

D. Polyatomic Ions

   1. Two common polyatomic cations are _____ and _____.

   2. Polyatomic anions usually have one or more oxygen atoms. These are

   collectively called _____.

   3. Memorize the names, formulas and charges of the polyatomic ions listed in Table 2.2 of your text.

E. Forming ionic compounds

   1. Rules

   a. Compounds have electrical neutrality. Na^+ and S^{2-} must be written as Na_2S, since you need two positive charges to balance the -2 charge on the S.

   b. The positive ion is always written before the negative ion.

   c. If two or more polyatomic ions are used in the formula, enclose the polyatomic ion in parentheses and put the number of ions you need outside the parentheses as a subscript.

   d. Do not write the charge of the ion in the formula. For example, sodium sulfide is Na_2S, not $Na^{2+}S^{2-}$; $2Na^+S^{2-}$; or $Na_2^+S^{2-}$.

2. *Exercises*

Predict the formulas of the ionic compounds formed by a

a. calcium ion and phosphate ion

Solution: Calcium ion is Ca^{2+} and phosphate ion is PO_4^{3-}. To balance the charges, we need three ions of Ca to get a total positive charge of 6, and two ions of phosphate to obtain a total negative charge of 6. Thus we write $Ca_3(PO_4)_2$; not $Ca_3P_2O_8$ or $Ca_3^{2+}(PO_4)_2^{3-}$.

b. manganese ion and bromine ion **(E6)**

c. sodium ion and nitrogen ion **(E7)**

d. aluminum ion and sulfate ion **(E8)**

VII. Names/Formulas of Compounds

A. Ionic compounds

   1. Kinds

     a. metal ion + nonmetal ion

      Example: _____

     b. metal ion + polyatomic ion

      Example: _____

     c. ammonium ion + nonmetal ion

      Example: _____

     d. ammonium ion + polyatomic ion

      Example: _____

   2. Rules

     a. When a metal is involved, the name of the metal is used.

     b. When the metal ion can have two different charges, the charge of the ion is indicated by writing it in Roman numerals in parentheses after the name of the metal.

      Example: Cu^+ is written as copper(I),

                Cu^{2+} is written as copper(II).

     c. When a nonmetal is involved, *ide* is added as a suffix to the root word of the nonmetal (usually the first syllable).

      Example: bromine becomes bromide

     d. Polyatomic ions retain their names.

    e. Exercises:
      Name the following compounds:
      (1) Li_3N **(E9)**

      (2) $(NH_4)_2CO_3$ **(E10)**

      (3) MgO **(E11)**

      (4) FeO
        Solution: Since oxygen always has a charge of 2-, in order to have electrical neutrality, iron has to have a charge of 2+. Thus, the compound's name is iron(II) oxide.
      (5) CuS **(E12)**

      (6) Fe_2S_3 **(E13)**

3. Most polyatomic ions contain one or more oxygen atoms. Collectively, these polyatomic ions are called oxoanions.
    a. Rules for naming oxoanions
      (1) The suffix *ate* is used _____

        _____

      (2) The suffix *ite* is used _____

        _____

      (3) The prefix *hypo* is used _____

        _____

      (4) The prefix *per* is used _____

        _____

    b. Memorize Table 2.3 in your text.
4. Writing the formula given the name of the compound
    Exercises: Write the formulas of the following compounds:

    a. sodium nitrite: **(E14)** _____

    b. rubidium dichromate: **(E15)** _____

    c. ammonium sulfate: **(E16)** _____

    d. calcium perchlorate: **(E17)** _____

B. Binary Molecular Compounds

  1. Rule: The first nonmetal gets its full name. The second nonmetal gets its root word + *ide*. Both nonmetals get a prefix denoting how many atoms are used to make the compound. However, when only one atom is used, the prefix mono is not attached.

    Example: CO_2 – first nonmetal = carbon; second nonmetal = dioxide

  2. Prefixes (Table 2.4)

    2 atoms - di

    3 atoms - _____

    4 atoms - _____

    5 atoms - _____

    6 atoms - _____

    7 atoms - _____

  3. *Exercises:*

    Name the following compounds:

    a. BrI **(E18)**

    b. PBr_3 **(E19)**

    c. SF_6 **(E20)**

  4. Acids

    a. Definition: _____

    b. When the gaseous binary molecules form acidic aqueous solutions, the *ide* suffix is changed to *ic*.

    c. When a hydrogen ion combines with an oxoanion, an oxoacid is formed. The prefixes of the oxoanion, *per* and *hypo*, remain unchanged. The suffix *ate* is changed to *ic*, while the suffix *ite* is changed to *ous*.

    Example: $HClO_3$ is derived from H^+ and ClO_3^- (chlorate). The name of the acid is chloric acid.

SELF-TEST
A. Multiple choice:
1. Fluorine has the nuclear symbol $^{19}_{9}$ F. The total number of subatomic particles (neutrons + protons + electrons) in a fluorine atom is
   a. 9   **b.** 10   **c.** 18   **d.** 19   **e.** 28

2. The combination of numbers and symbols in 5 N_2 stands for
   a. 5 nitrogen atoms   **b.** 10 separate nitrogen atoms
   c. 5 nitrogen molecules   **d.** 10 nitrogen molecules
   e. none of those

3. The number of protons in a molecule of CO_2 is
   a. 3   **b.** 12   **c.** 16   **d.** 22   **e.** 32

4. The formulas of ammonium sulfide and aluminum fluoride are
   a. NH_4SO_3 , Al_3F **b.** NH_4SO_4 , AlF_3
   c. $(NH_3)_2SO_3$, AlF_2 **d.** $(NH_4)_2S$, AlF_3
   e. $(NH_4)_2SO_3$, AlF_3

5. The law of multiple proportions can be applied to
   a. $^{1}_{1}$ H and $^{2}_{1}$ H   **b.** Cl and Cl_2
   c. CO_2 and SiO_2 **d.** CuO and Cu_2O

6. The formula of chromium(III) oxide is
   a. CrO   **b.** Cr_3O **c.** CrO_3 **d.** Cr_3O_2 **e.** Cr_2O_3

7. To form Fe^{3+}
   a. an Fe atom must gain 3 protons.
   b. an Fe^{2+} ion must gain 1 proton.
   c. an Fe atom must gain 3 electrons.
   d. an Fe^{2+} ion must gain 1 electron.
   e. none of the above will work.

8. Consider the following statements
   — There are always more neutrons than protons in an atom's nucleus.
   — The nucleus of any atom is heavier than all its electrons.
   — In ions, the number of protons is always greater than the number of electrons.
   — An atom has always more subatomic particles than an ion.
   The number of true statements is
   a. 0   **b.** 1   **c.** 2   **d.** 3   **e.** 4

9. In a chemical reaction, if an atomic particle carries a positive charge, then the atom has
   a. lost electrons   **b.** lost protons   **c.** lost neutrons
   d. gained electrons   **e.** gained protons

10. A nuclear symbol for the element simultaneously in group 5 and period 4 is

    a. $_{40}^{91}$Zr   **b.** $_{23}^{51}$V   **c.** $_{4}^{9}$Be   **d.** $_{5}^{11}$B

B. True or False:

_____ **1.** The electron has all the properties of the element.

_____ **2.** Neutrons have no charge and no mass.

_____ **3.** When an atom loses electrons, it becomes positively charged.

_____ **4.** Alpha (α) particles are identical to helium atoms.

_____ **5.** The simplest formula for $C_6H_{12}O_6$ is CH_2O.

_____ **6.** For the compound lithium chloride, the chloride ions are cations.

_____ **7.** Uncharged isotopes of the same element always have the same number of electrons.

_____ **8.** The proton is found in the nucleus of the atom.

_____ **9.** The mass number of an element represents the number of neutrons in the nucleus of that element.

_____ **10.** P–32 and S–32 have the same neutron to proton (n/p$^+$) ratio.

C. Fill in the blanks:

1. _____ $_{12}^{25}$Mg^{2+} has __(1)__ protons, __(2)__ neutrons and __(3)__ electrons.

2. _____

3. _____

4. _____   Period __(4)__ has no metals.

5. _____   The symbol for the metalloid in Group 13 is __(5)__.

6. _____   The name of the group to which sodium belongs is __(6)__.

7. _____   The name of the group to which neon belongs
                                is   (7) .

8. _____   The name of the halogen with 35 protons in its
                                nucleus is   (8) . (Write the name, not the sym-
                                bol.)

9. _____   If a new alkaline earth were created, its atomic
                                number would most probably be   (9) .

10. _____   The symbol for a metal that has 36 electrons
                                and whose cation has a +2 charge is   (10) .

11. _____   (11)   is the name of a post transition metal in
                                Period 4. (Write the name, not the symbol.)

12. _____   (12)   is the name of a molecule made up of
                                two atoms of an element with 7 protons and
                                three atoms of an element in Group 16, Period
                                2. (Write the name, not the formula.)

D. Problems: Consider the element zinc (Zn).
   1. Look it up in the Periodic Table. What is its atomic number?

   2. One of zinc's isotopes has a mass number of 64. Write the nuclear symbol for this isotope. Its ion is Zn^{2+}. How many electrons does its ion have?

   3. Describe zinc as a metal, nonmetal, or metalloid; main group, transition or post-transition element. Describe its place in the periodic table (group and period).

   4. Write the formula for the ionic compound formed when zinc ions (Zn^{2+}) combine with iodide ions (I^-).

5. Write the formulas of the following compounds:
   carbon disulfide
   aluminum carbonate
   perchloric acid
   ammonium permanganate
   cobalt(II) sulfide
   calcium hydride
   ammonia
   dinitrogen tetraoxide
   hydroiodic acid
   magnesium hypobromite

6. Write the names of the following compounds:
   $Na_2Cr_2O_7$
   P_4O_{10}
   $Fe_2(SO_4)_3$
   Na_2Se
   $Ca(HCO_3)_2$
   XeF_4
   CII_3
   CH_4
   H_2SO_3
   $Al(IO_2)_3$

ANSWERS
Exercises:
(E1) 13 protons, 14 neutrons, 13 electrons   **(E2)** CCl_4
(E3) $C_2H_4O_2$ **(E4)** 13 protons, 10 electrons
(E5) 16 protons, 18 electrons   **(E6)** $MnBr_2$
(E7) Na_3N **(E8)** $Al_2(SO_4)_3$ **(E9)** lithium nitride
(E10) ammonium carbonate **(E11)** magnesium oxide **(E12)** copper(II) sulfide
(E13) iron(III) sulfide   **(E14)** $NaNO_2$ **(E15)** $Rb_2Cr_2O_7$
(E16) $(NH_4)_2SO_4$ **(E17)** $Ca(ClO_4)_2$ **(E18)** bromine iodide
(E19) phosphorus tribromide **(E20)** sulfur hexafluoride

SELF-TEST
A. Multiple choice:
   1. e   **2.** c   **3.** d   **4.** d   **5.** d   **6.** e   **7.** e   **8.** b   **9.** a   **10.** b

B. True or False:
   1. F   **2.** F   **3.** T   **4.** F   **5.** T   **6.** F   **7.** T   **8.** T   **9.** F   **10.** F

C. Fill in the blanks:

1. 12	**2.** 13	**3.** 10	**4.** 1
5. B	**6.** alkali metal	**7.** noble gases	**8.** bromine
9. 120	**10.** Sr^{2+}	**11.** gallium	**12.** dinitrogen trioxide

D. Problems

1. 30

2. $^{64}_{30}$ Zn; 28

3. Zinc is a transition metal in group 12, period 4.

4. ZnI_2

5. CS_2 , $Al_2(CO_3)_3$, $HClO_4$, NH_4MnO_4 , CoS , CaH_2 , NH_3 , N_2O_4 , HI , $Mg(BrO)_2$

6. sodium dichromate, tetraphosphorus decaoxide, iron(III) sulfate, sodium selenide, calcium hydrogen carbonate, xenon tetrafluoride, chlorine triiodide, methane, sulfurous acid, aluminum iodite

Mass Relations in Chemistry; Stoichiometry

I. Mass

  A. Atomic mass

    1. Atomic masses are relative, representing the mass of an atom of one element compared to the mass of another. We use carbon-12 as the standard and assign it a mass of exactly 12 amu.

    2. Exercises

      a. Using the atomic masses in the Periodic Table, find an element that is four times heavier than carbon.

      Solution: Since carbon has an atomic mass of 12 amu, the element must have an atomic mass of about 48 amu. The Periodic Table lists Ti (titanium) with an atomic mass of 47.90 amu.

      b. Germanium (Ge) is about six times heavier than carbon. What is its approximate atomic mass? **(E1)**

      c. If helium (He) was the standard and assigned an atomic mass of 10 amu, what would the atomic mass of carbon be? **(E2)**

3. Experimental determination of relative atomic mass
  a. Instrument used: _____
  b. Describe briefly how the instrument determines the relative atomic mass.

4. Atomic mass and isotopic abundance
  a. Mathematical relationship

  b. Exercises
    (1) Antimony occurs in nature as a mixture of two isotopes: $^{123}_{51}$ Sb and
    $^{121}_{51}$ Sb. The isotope $^{121}_{51}$ Sb has a mass of 120.9 amu and an abundance
    of 57.25%. Antimony has an average atomic mass of 121.75 amu. What
    is the abundance of the other isotope? What is its atomic mass?
    Solution: Since there are only two isotopes, and the sum of their abun-
    dances has to be 100%, we have

$$100\% = \% \,^{121}_{51} \text{Sb} + \% \,^{123}_{51} \text{Sb}$$
$$^{123}_{51} \text{Sb} = 100.0\% - 57.25\%$$
$$= 42.75\%$$

The atomic mass (A.M.) of $^{123}_{51}$ Sb can be obtained from the formula

$$\text{A.M. } Y = (\text{A.M. } Y_1)(\%Y_1/100) + (\text{A.M. } Y_2)(\%Y_2/100)$$

where

$$\text{A.M. } Y = 121.75 \text{ amu}$$
$$\text{A.M. } Y_1 = 120.9 \text{ amu}$$
$$\frac{\% Y_1}{100} = 0.5725$$
$$\frac{\% Y_2}{100} = 0.4275$$

Thus,

$$121.75 = (120.9)(0.5725) + (A.M. \ Y_2)(0.4275)$$

$$A.M. \ Y_2 = \frac{121.75 - (120.9)(0.5725)}{0.4275}$$

$$= 122.9 \ amu$$

The atomic mass of $^{121}_{51}$ Sb is 122.9 amu.

(2) Silicon consists of three stable isotopes whose atomic masses and abundances are as follows:

$^{28}_{14}$ Si	27.98 amu	92.2%
$^{29}_{14}$ Si	28.98 amu	4.7%
$^{30}_{14}$ Si	29.97 amu	3.1%

Calculate the average atomic mass of Si. **(E3)**

B. Mass of a single atom

   1. Avogadro's number

      What does it represent? _____

      _____

   2. Exercises

     a. Calculate the mass of a carbon atom.

       Solution: We use the conversion factor: 6.022×10^{23} C atoms in 12.01 g of C. We can write this in the form

$$\frac{6.022 \times 10^{23} \ C \ atoms}{12.01 \ g \ C}$$

or

$$\frac{12.01 \ g \ C}{6.022 \times 10^{23} \ C \ atoms}$$

depending on which unit we want to cancel. In this exercise we want to eliminate the unit C atom and obtain the unit grams of C. Thus, we have

$$1 \text{ C atom} \times \frac{12.01 \text{ g C}}{6.022 \times 10^{23} \text{ C atoms}} = 1.994 \times 10^{-23} \text{g}$$

Note that atoms have extremely small masses. Check your answer to make sure you do *not* write 1.994×10^{23} g, which is a *huge* mass.

b. How many atoms are there in 2.483 g of C? **(E4)**

C. Mass of a molecule
  1. Formula mass

    Definition: _____
  2. Exercises
    a. What is the formula mass of chloric acid?
    Solution: Since formula mass means the sum of the atomic masses that make up the formula, we must first write the formula for chloric acid. (Review Chapter 2.) It is $HClO_3$. It is made up of one H atom (1.008 amu), 1 Cl atom (35.45 amu) and three O atoms (3×16.00 amu). The sum of these atomic masses is 84.46 amu.
    b. What is the formula mass for sodium nitrite? **(E5)**

  3. Relation of Avogadro's number to formula masses.
    A formula mass in grams has 6.022×10^{23} formula units.

II. Mole
  A. Definition: _____

  _____

  B. Molar masses
    1. Definition: _____

    2. Symbol: _____
    3. *Exercises*
      a. Calculate the molar mass of Al_2O_3 in grams/mol.
      Solution: We first obtain the formula mass of aluminum oxide, Al_2O_3. Remember that there are two atoms of aluminum and three atoms of oxygen. Thus the formula mass is

$$2(26.98 \text{ amu}) + 3(16.00 \text{ amu}) = 101.96 \text{ amu}$$

The molar mass of Al_2O_3 is then obtained by substituting the units g/mol for the unit amu. The molar mass ($\mathcal{M}$) of Al_2O_3 is 101.96 g/mol.

    b. Calculate the molar masses of

       (1) ammonium sulfate — $(NH_4)_2SO_4$ **(E6)**

       (2) aspartic acid — $C_4H_7NO_4$ **(E7)**

C. *Mole-Gram Conversion Exercises*

    1. How many moles are there in 1.00 g of Al_2O_3?

      Solution: The preceding worked exercise gave us the molar mass of Al_2O_3: 101.96 g/mol. This can now be used as a conversion factor, that is,

$$\frac{101.96 \text{ g } Al_2O_3}{1 \text{ mol } Al_2O_3} \quad \text{or} \quad \frac{1 \text{ mol } Al_2O_3}{101.96 \text{ g } Al_2O_3}$$

depending on which unit we wish to eliminate. This exercise asks that we cancel the unit gram and obtain the unit mole. Thus, we obtain

$$1.00 \text{ g } Al_2O_3 \times \frac{1 \text{ mol } Al_2O_3}{101.96 \text{ g } Al_2O_3} = 0.00981 \text{ mol}$$

Remember, before you can convert grams to moles or vice versa, you must have the conversion factor, which is simply the formula mass, in grams, of one mole of the substance with which you are working.

    2. Find the number of grams in 3.874 mol of NH_4Cl. **(E8)**

    3. Find the number of moles in 13.70 g of $Ba(OH)_2$. **(E9)**

III. **Mass Relations in Chemical Formulas**

    A. What a chemical formula can tell you.

      1. The atom ratios of the different elements in the compound.

        Example: NH_4Cl — 1 atom N : 4 atoms H : 1 atom Cl

      2. The mole ratios of the different elements in the compound.

        Example: NH_4Cl — 1 mol N : 4 mol H : 1 mol Cl

3. The mass ratios of the different elements in the compound.
Example: NH_4Cl — 14.01 g N : 4(1.01) g H : 35.45 g Cl

B. Percent Composition from Formula

1. Strategy: To get mass percent you will need to know:

a. the mass of one mole of element

b. the total mass of each element (quantity from **(a)** × number of atoms of the element in the formula)

c. the total mass of the compound (sum of the masses obtained from **(b)**)

To calculate mass percent, use the formula

$$\text{percent element} = \frac{\text{total mass of element}}{\text{total mass of compound}} \times 100$$

Remember: When calculating percent composition, check your answer to see whether the total of the percent masses of all atoms is about 100%. If you get a total percent of 25.3% or one that is 263%, something is dreadfully wrong!

2. *Exercises*

a. Calculate the mass percent of each element in Al_2O_3. How much aluminum (Al) can be obtained from 1.000 kg of Al_2O_3?

Solution: For this exercise

(1) the elements are Al and O, and the mass of 1 mol of Al is 26.98 g, whereas the mass of 1 mol of O is 16.00 g.

(2) Total mass of Al = 26.98 × 2 = 53.96
Total mass of O = 16.00 × 3 = 48.00

(3) We now can obtain the mass percent of each atom.

$$\text{Total mass of compound} = \text{mass of Al} + \text{mass of O}$$
$$= 53.96 + 48.00$$
$$= 101.96 \, g$$

$$\text{Mass \% Al} = \frac{53.96}{101.96} \times 100 = 52.92 \,\%$$

$$\text{Mass \% O} = \frac{48.00}{101.96} \times 100 = 47.08 \,\%$$

Check: 52.92 + 47.08 = 100%. The answer is reasonable.

For the second part of the exercise, you have to convert 1.000 kg of Al_2O_3 to g of Al. This can be easily done by using the conversion factor:

$$\frac{53.96 \, g \, Al}{101.96 \, g \, Al_2O_3}$$

Thus,

$$1000 \, g \, Al_2O_3 \times \frac{53.96 \, g \, Al}{101.96 \, g \, Al_2O_3} = 529.2 \, g \, Al$$

As you can see, the second part of the exercise can be done without knowing the percent composition of Al_2O_3. The formula suffices.

b. Calculate the mass percent of each element in cholesterol. Its simplest formula is $C_{27}H_{46}O$. **(E10)**

c. Freon-12, a refrigerant, is made up of 1 carbon atom, 2 chlorine atoms, and 2 fluorine atoms. Write the formula for Freon-12. How many grams of fluorine are required to make 100.0 g of Freon-12? **(E11)**

C. Simplest Formula from Chemical Analysis

1. Steps in determining the simplest formula

   a. Determine the mass of each element.

   b. Determine the number of moles of each element by taking the quantity in Step **(a)** and dividing by the atomic mass of the element.

   c. Pick out the smallest number of moles in Step **(b)**.

   d. Divide the number of moles of each element (Step **(b)**) by the smallest number of moles (Step **(c)**).

   e. The answers to Step **(d)** are usually integers and thus are used as subscripts.

2. *Exercises*

   a. A sample of compound is made up of 78.20 g of potassium (K) and 32.06 g of sulfur (S). What is its simplest formula?

Solution: Step (a) is given:

$$78.20 \text{ g K}; \quad 32.06 \text{ g S}$$

Step (b):

$$78.20 \text{ g K} \times \frac{1 \text{ mol K}}{39.10 \text{ g K}} = 2.000 \text{ mol K}$$

$$32.06 \text{ g S} \times \frac{1 \text{ mol S}}{32.06 \text{ g S}} = 1.000 \text{ mol S}$$

Step (c): Smallest number of moles = 1
Step (d):

$$\text{mol ratio K} = \frac{2.000}{1.000} = 2$$

$$\text{mol ratio S} = \frac{1.000}{1.000} = 1$$

Step (e): K_2S

  b. A compound is made up of 11.80 g of sulfur (S), 16.92 g of sodium (Na), and 23.55 g of oxygen (O). What is its simplest formula? **(E12)**

3. Simplest formula from percent composition:
   Steps (a) through (e) are used. For convenience, assume 100.0 g to obtain quantities needed for Step (a).
   Exercises:
   a. A compound has the following percent composition:

$$31.9 \% \text{ K}; \qquad 29.0 \% \text{ Cl}; \qquad \text{and} \qquad 39.1 \% \text{ O}$$

Determine its simplest formula.
   Solution:
   Step (a): To obtain the mass of each element, we assume 100.0 g of compound. Thus, K = 31.9 g, Cl = 29.0 g, O = 39.1 g.

Step (b): Moles of each element

$$K: \quad 31.9 \text{ g} \times \frac{1 \text{ mol}}{39.10 \text{ g}} = 0.816 \text{ mol}$$

$$Cl: \quad 29.0 \text{ g} \times \frac{1 \text{ mol}}{35.45 \text{ g}} = 0.818 \text{ mol}$$

$$O: \quad 39.1 \text{ g} \times \frac{1 \text{ mol}}{16.00 \text{ g}} = 2.44 \text{ mol}$$

Step (c): Smallest number of moles = 0.816
Step (d): Ratios

$$K: \quad \frac{0.816}{0.816} = 1.00$$

$$Cl: \quad \frac{0.818}{0.816} = 1.00$$

$$O: \quad \frac{2.45}{0.816} = 2.99$$

Step (e): $K_1Cl_1O_3$
The simplest formula is $KClO_3$

A word about rounding off: In determining integer ratios, one simply cannot round off a three-significant-figure ratio to a one-significant-figure ratio. To get an integer, it is permissible to round down by at most 0.1. Thus, 2.09 can be rounded down to 2. But 2.23 cannot be rounded down to 2. It is likewise permissible to round up by as much as 0.1. Thus, 2.93 can be rounded up to 3, whereas 2.85 cannot be rounded up to 3. What does one do when Step (d) does not yield whole integers? Multiply each ratio by small whole numbers until multiplication by one small number yields ratios that are all in integers.

b. A compound is made up of C, Cl, and O atoms. It has 12.13% of C and 70.91% of Cl by mass. What is its empirical formula? **(E13)**

4. Simplest formula given raw data from experimental analysis
   The results of an experimental analysis to determine the simplest formula of a compound can be given in terms of masses of compounds. Hence, Step (a) has to be modified. You have to determine the mass of the element desired from the given mass of compound obtained in the analysis. Steps (b) through (e) are the same as before.

   Exercises

   a. A compound with mass 1.083 g contains only carbon and sulfur. It is burned in oxygen. The mass of CO_2 obtained after burning is 0.627 g. What is the simplest formula of the compound?

   Solution:

   Step (a): Mass of the elements
   In this exercise, the mass of carbon is given in terms of CO_2. We therefore need a conversion factor to change g of CO_2 to g of C.

$$1 \text{ mol } CO_2 \text{ contains } 1 \text{ mol C}$$
$$12.01 \text{ g C is present in } 44.01 \text{ g } CO_2.$$

   The conversion factor is

$$\frac{44.01 \text{ g } CO_2}{12.01 \text{ g C}} \quad \text{or} \quad \frac{12.01 \text{ g C}}{44.01 \text{ g } CO_2}$$

   We can now convert grams of CO_2 to grams of C. We have

$$0.627 \text{ g } CO_2 \times \frac{12.01 \text{ g C}}{44.01 \text{ g } CO_2} = 0.171 \text{ g}$$

   Since the compound contains only C and S, we can write
$$\text{Mass of S} = 1.083 - 0.171 = 0.912 \text{ g}$$

   Step (b): Moles of elements

$$\text{C}: \quad 0.171 \text{ g} \times \frac{1 \text{ mol}}{12.01 \text{ g}} = 0.0142 \text{ mol C}$$

$$\text{S}: \quad 0.912 \text{ g} \times \frac{1 \text{ mol}}{32.07 \text{ g}} = 0.0284 \text{ mol S}$$

   Step (c): Smallest number of moles = 0.0142
   Step (d): Mole ratios

$$\text{C}: \quad \frac{0.0142}{0.0142} = 1.00$$

$$\text{S}: \quad \frac{0.0284}{0.0142} = 2.00$$

   Step (e): C_1S_2
   The simplest formula of the compound is CS_2

   b. An unknown compound contains only carbon, hydrogen, and oxygen. If 1.500 g of the compound is burned with an excess of oxygen, 1.433 g

CO_2 and 0.582 g H_2O are produced. What is the simplest formula of the compound? **(E14)**

D. Molecular Formula from the Simplest Formula

To determine the molecular formula from the simplest formula, one has to know the molar mass of the compound. Since the molecular formula is a whole number multiple of the simplest formula, it follows that

$$\text{Multiple} = \frac{\text{molar mass}}{\text{simplest formula mass}}$$

Here, simplest formula mass is the sum of the atomic masses of the elements that make up the simplest formula. The subscripts in the simplest formula are multiplied by the multiple to get the subscripts for the molecular formula.

Exercises

1. Propene has the simplest formula CH_2. What is its molecular formula if it has a molar mass of 42.09 g/mol?

Solution:

$$\text{simplest formula mass} = 12.00 + 2(1.008) = 14.02 \ \frac{\text{g}}{\text{mol}}$$

$$\text{Multiple} = \frac{42.09}{14.02} = 3$$

Thus,

$$\text{Molecular formula} = C_{1\times3}H_{2\times3} = C_3H_6$$

2. Determine the molecular formula of ethylene glycol. It has an atom ratio of 1 C : 1 O : 3 H. Its molar mass is 62.06 g/mol. **(E15)**

IV. Reactions
A. Writing and Balancing Equations
1. Steps in writing a balanced equation:
   a. _____
      _____
   b. _____
      _____
   c. _____
      _____

2. Points to remember
   a. Balance equations by adjusting coefficients in front of the formulas, never by changing the subscripts within the formulas.
   b. Start balancing with an element that appears only once on each side of the equation.
   c. All metals at room temperature are solids except Hg, which is a liquid.
   d. All ionic compounds (metal + nonmetal) are solids.
   e. You must know the names and the ionic charges of the elements and compounds in order to be able to write equations correctly (Chapter 2).

3. *Exercises*
   a. Write a balanced equation for the reaction between sodium and chlorine to form sodium chloride.
      Solution:
      (1) Recall sodium is written as Na and as an ion has a +1 charge. Also, recall chlorine is written Cl and is diatomic: thus, Cl_2. Since it belongs to Group **17**, its charge as an ion is −1. Thus,
$$Na + Cl_2 \rightarrow NaCl$$

      (2) We need two Cl atoms in the product side to balance the two atoms on the reactant side.
$$Na + Cl_2 \rightarrow 2\,NaCl$$

(3) Now we need two atoms of Na in the reactant side to balance the two atoms of Na on the product side.

$$2\,Na + Cl_2 \rightarrow 2\,NaCl$$

(4) Sodium is a solid; chlorine is a gas; and sodium chloride, an ionic compound, is a solid.

$$2\,Na\,(s) + Cl_2\,(g) \rightarrow 2\,NaCl\,(s)$$

b. Write a balanced equation for
   (1) the reaction of chromium with sulfur. **(E16)**

   (2) the reaction of bismuth with oxygen. **(E17)**

B. Mass Relations from Equations
   1. Mole-mole relationships
   a. The coefficients indicate how many moles of one reactant are required to combine with another reactant. They also show how many moles of product are obtained.
   b. These coefficients can be used as conversion factors for the particular equation you are dealing with.
   c. Exercises
      (1) Consider the reaction between methane gas (CH_4) and oxygen gas.

$$CH_4\,(g) + 2\,O_2\,(g) \rightarrow CO_2\,(g) + 2\,H_2O\,(\ell)$$

How many moles of oxygen are required to react with 3.87 mol of CH_4? *Solution:* This equation can be written as

$$1\text{ mol } CH_4 + 2\text{ mol } O_2 \rightarrow 1\text{ mol } CO_2 + 2\text{ mol } H_2O$$

The quantities 1 mol CH_4, 2 mol O_2, 1 mol CO_2, and 2 mol H_2O are chemically equivalent to each other in this reaction. Conversion factors can be

$$\frac{2\text{ mol } O_2}{1\text{ mol } CO_2} \quad ; \quad \frac{1\text{ mol } CH_4}{2\text{ mol } H_2O}$$

or a variety of other combinations. Since CH_4 and O_2 are involved, the conversion factor that may be used is either

$$\frac{2 \text{ mol } O_2}{1 \text{ mol } CH_4} \quad \text{or} \quad \frac{1 \text{ mol } CH_4}{2 \text{ mol } O_2}$$

Now it is simply a matter of converting moles of CH_4 to moles of O_2.

$$3.87 \text{ mol } CH_4 \times \frac{2 \text{ mol } O_2}{1 \text{ mol } CH_4} = 7.74 \text{ mol}$$

Remember: You will not get the right conversion factors if the equation is not balanced correctly.

(2) How many moles of hydrogen gas are required to produce 1.76 mol of iron in the reaction

$$Fe_2O_3 \text{ (s)} + 3\,H_2 \text{ (g)} \;\rightarrow\; 2\,Fe \text{ (s)} + 3\,H_2O \text{ (}\ell\text{)} \qquad \textbf{(E18)}$$

2. Gram-mole relationships

Once the mole-mole relationship is established, other conversion factors can be generated by using the fact that one mole of any substance is equivalent to its gram molar mass.

Exercises

a. Consider the reaction between potassium and iodine to form potassium iodide. How many grams of iodine are necessary to produce 6.97 mol of product?

Solution:

(1) First, you have to write a balanced equation.

$$2\,K \text{ (s)} + I_2 \text{ (s)} \;\rightarrow\; 2\,KI \text{ (s)}$$

(2) Next, write the mole-mole relationships.

$$2 \text{ mol } K + 1 \text{ mol } I_2 \;\rightarrow\; 2 \text{ mol } KI$$

(3) Now write the gram-mole relationship of each element or compound.

$$2 \text{ mol } K = 39.10 \;\frac{g}{mol} \times 2 \text{ mol} = 78.20 \text{ g}$$

$$1 \text{ mol } I_2 = 126.9 \;\frac{g}{mol} \times 2 \text{ mol} = 253.8 \text{ g}$$

(I_2 is diatomic so there are two mol I per mol of I_2.)

$$2 \text{ mol } KI = 166.0 \;\frac{g}{mol} \times 2 \text{ mol} = 332.0 \text{ g}$$

(4) Write the gram-mole relationships:

$$2 \text{ mol } (78.20 \text{ g}) \text{ K} + 1 \text{ mol } (253.8 \text{ g}) \text{ I}_2 \rightarrow 2 \text{ mol } (332.0 \text{ g}) \text{ KI}$$

(5) Now, since you want to convert 6.87 mol of KI to grams I$_2$, pick the values for grams of I$_2$ per 2 mol of KI for your conversion factor. That will therefore be

$$\frac{253.8 \text{ g I}_2}{2 \text{ mol KI}} \quad \text{or} \quad \frac{2 \text{ mol KI}}{253.8 \text{ g I}_2}$$

(6) Thus, we have

$$6.87 \text{ mol KI} \times \frac{253.8 \text{ g I}_2}{2 \text{ mol KI}} = 872 \text{ g I}_2$$

b. The reaction between dinitrogen trioxide gas and liquid water yields an aqueous solution of nitrous acid, HNO_2 (aq).

(1) Write and balance the equation for the reaction.

(2) How many grams of nitrous acid would be made from 0.874 mol of dinitrogen trioxide gas?

(3) How many milliliters of water are needed to react with 0.874 mol of dinitrogen trioxide? (Density of water is 1.00 g/mL.) **(E19)**

3. Gram-gram relationships

You use the gram-mole relationship, except that you will now convert from grams of one substance to grams of another. You should follow all the steps in the gram-mole relationship problem.

Exercises

a. Consider again the reaction

$$2\,K\,(s) + I_2\,(s) \;\rightarrow\; 2\,KI\,(s)$$

How many grams of KI are produced by reacting 6.029 g of K?

Solution:

(1) The gram-mole relationships are the same as those for the last worked exercise.

$$2 \text{ mol } (78.20\,g)\,K + 1 \text{ mol } (253.8\,g)\,I_2 \;\rightarrow\; 2 \text{ mol } (332.0\,g)\,KI$$

(2) Now pick a conversion factor for this problem. Since you want to relate grams of K to grams of KI, you should use

$$\frac{332.0\,g\,KI}{78.20\,g\,K} \qquad \text{or} \qquad \frac{78.20\,g\,K}{332.0\,g\,KI}$$

Thus,

$$6.029\,g\,K \times \frac{332.0\,g\,KI}{78.20\,g\,K} = 25.60\,g\,KI$$

With more practice and confidence, you will not need to write out the whole string of relationships as previously shown. You will be able to pick out quickly what you need without writing all of them down. For now, however, you should proceed in this manner.

b. When aluminum reacts with lead(IV) oxide (PbO_2), the products are aluminum oxide and lead.

(1) Write a balanced equation for the reaction.

(2) How many grams of aluminum are required to produce 12.17 g of lead?
(E20)

C. Limiting Reactant and Theoretical Yield
  1. Definitions
     a. Limiting reactant

     _____

     _____

     b. Theoretical yield

     _____

     _____

  2. Steps to be followed in determining the limiting reactant of a reaction

     a. _____

     _____

     b. _____

     _____

     _____

     c. _____

     _____

  3. *Exercises*
     a. When hydrazine (N_2H_4) reacts with hydrogen peroxide (H_2O_2), the products are nitrogen gas and liquid water. If 2.686 g of hydrazine reacts with 3.143 g of H_2O_2, how many grams of nitrogen gas are produced? What is the limiting reactant?
     Solution: First, we write a balanced equation.

$$N_2H_4 \, (\ell) + 2\,H_2O_2 \, (\ell) \; \rightarrow \; 4\,H_2O \, (\ell) + N_2 \, (g)$$

Step 1: Determine how much N_2 will be produced if all the hydrazine is consumed. That means converting grams of hydrazine into grams of nitrogen. The gram-mole relationships are

1 mol (32.06 g) N_2H_4 + 2 mol (68.04 g) $H_2O_2 \; \rightarrow$

4 mol (72.08 g) H_2O + 1 mol (28.02 g) N_2

The conversion factor to use is

$$\frac{28.02 \text{ g } N_2}{32.06 \text{ g } N_2H_4}$$

We thus obtain the mass of nitrogen gas produced by consuming all the hydrazine:

$$2.686 \text{ g } N_2H_4 \times \frac{28.02 \text{ g } N_2}{32.06 \text{ g } N_2H_4} = 2.348 \text{ g } N_2$$

Step 2: Determine how much N_2 will be produced if all the H_2O_2 is consumed; that is, convert grams of H_2O_2 to grams of N_2. The conversion factor to use in this case is

$$\frac{28.02 \text{ g } N_2}{68.04 \text{ g } H_2O_2}$$

Thus the mass of nitrogen gas obtained if all the H_2O_2 is consumed is

$$3.143 \text{ g } H_2O_2 \times \frac{28.02 \text{ g } N_2}{68.04 \text{ g } H_2O_2} = 1.294 \text{ g } N_2$$

Step 3: The smaller quantity is 1.294 g N_2, which is the *theoretical yield*. The reactant producing that quantity, in this case H_2O_2, is the limiting reactant.

b. How many grams of hydrazine in Exercise (*a*) are left over?

Solution: To determine this, you must first calculate how much hydrazine really reacted. Exercise (*a*) shows that H_2O_2 and not N_2H_4 was completely consumed. To determine then how much N_2H_4 is required to react with H_2O_2, the limiting reactant, you have to convert grams of H_2O_2 to grams of N_2H_4.

Caution: It will not work if you convert grams of N_2H_4 to grams of H_2O_2. You have to take the mass of the *limiting reactant* and convert it to the grams of the excess reactant (the nonlimiting reactant).

The conversion factor is

$$\frac{32.06 \text{ g } N_2H_4}{68.04 \text{ g } H_2O_2}$$

Hence we get

$$3.143 \text{ g } H_2O_2 \times \frac{32.06 \text{ g } N_2H_4}{68.04 \text{ g } H_2O_2} = 1.481 \text{ g } N_2H_4$$

Since we had 2.686 g of hydrazine to start with and we reacted 1.481 g as shown above, then the amount of unreacted hydrazine is 1.205 g.

c. Consider the reaction

$$2 NH_3 \text{ (g)} + H_2S \text{ (g)} \rightarrow (NH_4)_2S \text{ (s)}$$

If 6.837 g of NH_3 reacts with 4.129 g of H_2S:

(1) What is the theoretical yield of $(NH_4)_2S$?

(2) What is its limiting reactant?

(3) What is the reactant in excess and how many grams of that reactant are left over after the reaction has been completed? **(E21)**

Caution: If you get a negative answer, start all over again. Something is wrong!

D. Percent Yield

    1. Formula

_____

    2. Definition of actual yield: _____

_____

    3. Exercises

        a. When 1.26 g of nitrogen are combined with hydrogen, 0.874 g of ammonia are produced.

(1) What is the theoretical yield?

Solution: The balanced equation for the reaction is

$$N_2\ (g) + 3\,H_2\ (g) \;\rightarrow\; 2\,NH_3\ (g)$$

The theoretical yield is the result of the conversion of grams of N_2 to grams of NH_3. The gram-mole relationships are

$$1\ \text{mol}\ (28.02\,\text{g})\ N_2 + 3\ \text{mol}\ (6.048\,\text{g})\ H_2\ \rightarrow\ 2\ \text{mol}\ (34.07\,\text{g})NH_3$$

The conversion factor is

$$\frac{34.07\,\text{g}\ NH_3}{28.07\,\text{g}\ N_2}$$

Thus the mass of ammonia theoretically obtained from nitrogen is

$$1.26\,\text{g}\ N_2 \times \frac{34.07\,\text{g}\ NH_3}{28.02\,\text{g}\ N_2} = 1.53\,\text{g}\ NH_3$$

The theoretical yield is 1.53 g NH_3

(2) What is the actual yield?

Solution: The actual yield is the amount of NH_3 the problem says was produced, that is, 0.874 g.

(3) What is the percent yield?

Solution:

$$\text{Percent yield} = \frac{\text{actual yield}}{\text{theoretical yield}} \times 100$$

$$= \frac{0.874}{1.53} \times 100 = 57.1\%$$

b. When 1.417 g of carbon reacts with excess hydrogen, 1.67 g of ethane gas (C_2H_6) are produced.

(1) Write a balanced equation for the reaction.

(2) What is the theoretical yield?

(3) What is the percent yield? **(E22)**

SELF-TEST
A. Multiple choice:
1. Copper(II) oxide reacts with hydrogen gas, producing copper metal and water. The correct form of the chemical equation that describes this reaction is
   a. Cu_2O (s) + H_2 (g) → 2 Cu (s) + H_2O (ℓ)
   b. Cu_2O (s) + 2 H (g) → 2 Cu (s) + H_2O (ℓ)
   c. CuO (s) + H_2 (g) → Cu (s) + H_2O (ℓ)
   d. CuO (s) + 2 H (g) → Cu (s) + H_2O (ℓ)
   e. None of the above equations is correct.

2. The units for Avogadro's number can be
   a. molecules per mole.   **b.** atoms per mole.
   c. ions per mole.   **d.** all of the above.

3. How many kg of copper could one theoretically obtain from one kilogram of copper fluorosilicate, Cu_2SiF_6 ($\mathcal{M}$ = 269.1 g/mol)?
   a. 2(63.5)/1000   **b.** 2(63.5)(269.1)   **c.** 2(63.5)/269.1
   d. 2(63.5)(269.1)/1000   **e.** 2(63.5)(1000)/269.1

4. The atomic mass of vanadium is 50.942. Vanadium consists of two isotopes: $^{50}_{23}$V, mass 49.947 and $^{51}_{23}$V, mass 50.944. From this we can conclude that the abundance of $^{51}_{23}$V is

   a. about 10%   **b.** about 33%   **c.** about 66%   **d.** about 99%

5. Which one of the following particles has a mass that is closest to the mass of the oxygen atom?

   a. F atom   **b.** OH^- ion   **c.** O^{2-} ion   **d.** H_2O molecule

6. The number of molecules in a sample of NH_3 weighing 8.50 g is

   a. 0.500   **b.** 1.00   **c.** 144   **d.** 3.01×10^{23} **e.** 6.02×10^{23}

7. A compound contains 29.3% C, 57.7% Cl and 13.0% O by mass. Its molar mass is 246 g/mol. Its molecular formula is: (All choices have $\mathcal{M} = 246$ g/mol)

   a. $C_6Cl_4O_2$ **b.** $C_3Cl_5O_2$ **c.** $C_2Cl_4O_5$ **d.** $C_9Cl_3O_2$

8. Which of the following statements about caffeine, $C_8H_{10}O_2N_4$ ($\mathcal{M} = 194$ g/mol) is NOT correct?

   a. Four moles of caffeine have a mass of 776 g.

   b. The simplest formula for caffeine is $C_8H_{10}O_2N_4$.

   c. Caffeine contains 28.9% nitrogen.

   d. One mole of caffeine has 6.02×10^{24} hydrogen atoms.

9. How many selenium atoms are required to have the same mass as 126 atoms of sulfur?

   a. $\dfrac{126 \times 32.06}{78.96}$ **b.** $\dfrac{126 \times 78.96}{32.06}$

   c. $\dfrac{32.06}{126 \times 78.96}$ **d.** $\dfrac{78.96}{126 \times 32.06}$

10. The molecular formula of a substance

   a. is sometimes equal to its simplest formula.

   b. is always a whole number multiple of its simplest formula.

   c. indicates the exact number and the identities of the atoms that make up a molecule.

   d. has the same % composition as that of its simplest formula.

   e. a, b, and c are true.

   f. a, b, c, and d are true.

B. Fill in the blanks:

1. Consider a 5.871 g sample of pure nickel metal.

_____ **a.** How many moles of nickel are in this sample?

_____ **b.** How many atoms of nickel are in this sample?

_____ **c.** What is the ratio of nickel atoms to silver atoms in 5.781 g of nickel and 0.0108 g of silver?

2. Consider the balanced equation for the reaction in which xenon hexafluoride reacts with water forming xenon, oxygen and hydrogen fluoride gases.

$$2\,XeF_6\,(g) + 6\,H_2O\,(\ell) \rightarrow 2\,Xe\,(g) + 3\,O_2\,(g) + 12\,HF\,(g)$$

In an experiment, when 5.0 mol of xenon hexafluoride are made to react with an excess of water, 4.0 moles of xenon are obtained.

_____ **a.** What is the theoretical yield (in moles) for Xe when 5.0 moles of XeF_6 react?

_____ **b.** What is the theoretical yield (in grams) for oxygen when 5.0 moles of XeF_6 react?

_____ **c.** What is the percent yield of Xe in the experiment described above?

C. Problems: Consider the element zinc (Zn).

1. If zinc was designated to have an atomic mass of 100.0 amu, what would the atomic mass of Si be?

2. How many atoms would there be in 1.00 g of zinc? How many moles are there in 3.48 g of zinc? What is the mass of two million atoms of zinc?

3. Zinc has two isotopes. Zn-64 has an atomic mass of 63.9292 amu and Zn-72 has an atomic mass of 71.9278 amu. The average atomic mass for zinc is 65.38 amu. Calculate the abundance, in percent, of Zn-64.

Consider the molecule cyclohexene, which contains only hydrogen and carbon atoms.

4. When a sample is burned in oxygen, 4.822 g of CO_2 and 1.650 g of H_2O are obtained. What is its simplest formula?

5. Calculate the mass percent of carbon and hydrogen in cyclohexene.

6. How many grams of carbon would you obtain from 3.87 g of sample?

7. The molar mass of cyclohexene is 82.10 g/mol. What is its molecular formula?

8. Write and balance the equation for the reaction between liquid cyclohexene and oxygen forming carbon dioxide and water.

Consider the reaction

$$Mg_3N_2 \ (s) + H_2O \ (\ell) \ \rightarrow \ Mg(OH)_2 \ (s) + NH_3 \ (g)$$

9. Balance the equation and name each compound.

10. How many moles of $Mg(OH)_2$ are produced from 0.319 mol of Mg_3N_2?

11. How many grams of water are necessary to produce 4.39 mol of NH_3?

12. If 12.48 g of $Mg(OH)_2$ are formed, how many grams of NH_3 are produced?

13. Start with 2.95 g of magnesium nitride and 1.500 g of water.

    a. What is the limiting reactant?

    b. How many grams of magnesium hydroxide can be theoretically obtained?

    c. If only 2.009 g of magnesium hydroxide are obtained, what is the percent yield of the reaction?

ANSWERS
Exercises:

(E1) 72 amu   **(E2)** 30 amu   **(E3)** 28.1 amu   **(E4)** 1.245×10^{23} C atoms

(E5) 69.00 amu   **(E6)** 132.14 g/mol   **(E7)** 133.12 g/mol   **(E8)** 207.3 g

(E9) 0.07998 mol   **(E10)** 83.87% C; 11.99% H; 4.14% O   **(E11)** CCl_2F_2 ; 31.43 g

(E12) SNa_2O_4 **(E13)** CCl_2O **(E14)** CO_2H_2 **(E15)** $C_2O_2H_6$

(E16) $2\,Cr\ (s) + 3\,S\ (s) \rightarrow Cr_2S_3\ (s)$ **(E17)** $4\,Bi\ (s) + 3\,O_2\ (g) \rightarrow 2\,Bi_2O_3\ (s)$

(E18) 2.64 mol   **(E19)** $N_2O_3\ (g) + H_2O\ (\ell) \rightarrow 2\,HNO_2\ (aq)$; 82.2g; 15.7 mL

(E20) $4\,Al\ (s) + 3\,PbO_2\ (s) \rightarrow 2\,Al_2O_3\ (s) + 3\,Pb\ (s)$; 2.113 g

(E21) 8.258 g; H_2S; NH_3; 2.708 g

(E22) $2\,C\ (s) + 3\,H_2\ (g) \rightarrow C_2H_6\ (g)$; 1.774 g; 94.1%

Self-Test
A. Multiple Choice:

   1. c   **2.** d   **3.** c   **4.** d   **5.** c   **6.** d   **7.** a   **8.** b   **9.** a   **10.** f

B. Fill in the blanks:

   1. a. 0.100   **b.** 6.022×10^{22} **c.** $1.00 \times 10^3 : 1$

   2. a. 5.0   **b.** 2.4×10^2 **c.** $8.0 \times 10^1 \%$

C. Problems:

1. 42.97 amu   **2.** 9.20×10^{21} atoms; 0.0532; 2.17×10^{-16} g   **3.** 81.9%

4. C_3H_5 **5.** 87.6% C, 12.4% H   **6.** 3.39 g

7. C_6H_{10} **8.** $2\,C_6H_{10}\,(\ell) + 17\,O_2\,(g) \rightarrow 12\,CO_2\,(g) + 10\,H_2O\,(\ell)$

9. $Mg_3N_2\,(s) + 6\,H_2O\,(\ell) \rightarrow 3\,Mg(OH)_2\,(s) + 2\,NH_3\,(g)$ Mg_3N_2 = magnesium nitride, H_2O = water $Mg(OH)_2$ = magnesium hydroxide NH_3 = ammonia

10. 0.957 mol   **11.** 237 g   **12.** 2.430 g

13. a. H_2O **b.** 2.428 g   **c.** 82.74%

4

Reactions in Aqueous Solutions

I. Solute Concentration – Molarity

  A. Definition of molarity: _____

_____

  B. Units: _____

  C. Exercises:

    1. How would you prepare 250.0 mL of a 0.01269 M solution of $Ba(OH)_2$?

    Solution: The 0.01269 M means you want a solution that contains 0.01269 mol of $Ba(OH)_2$ per liter. Since the mole is not a unit that can be measured or counted, you will have to express it by using another unit that you can deal with in the laboratory. That unit is grams. You can weigh out $Ba(OH)_2$ in a balance that gives a mass in grams. Thus, you have to "convert" moles/liter to grams/liter. The conversion factor is the molar mass of $Ba(OH)_2$, which is 171.3 g/mol (or $\mathcal{M}$ = 171.3 g/mol). Hence we get

$$0.01269 \ \frac{\text{mol}}{\text{L}} \times \frac{171.3 \text{ g}}{1 \text{ mol}} = \frac{2.174 \text{ g}}{\text{L}}$$

However, you want to prepare only 250.0 mL of solution. Thus,

$$\frac{2.174 \text{ g}}{\text{L}} \times 0.2500 \text{ L} = 0.5435 \text{ g}$$

2. A student has 2.687 g of $(NH_4)_2SO_4$. What volume of solution does she need to obtain a solution that is 0.200 M? **(E1)**

3. What is the molarity of a solution of glucose ($C_6H_{12}O_6$) prepared by adding 126.3 g of glucose to a volumetric flask and dissolving it in enough water to make 500.0 mL of solution? **(E2)**

D. Molarity of ions in solution
   1. The concentration of an ion sometimes differs from that of the ionic compound it comes from.
   2. *Example:*
      When 1.00 mol of $Ba(NO_3)_2$ is dissolved in enough water to make one liter of solution, a 1.00 M solution of $Ba(NO_3)_2$ is obtained. However, when $Ba(NO_3)_2$ dissolves, the following takes place:

$$Ba(NO_3)_2 \text{ (s)} \rightarrow Ba^{2+} \text{ (aq)} + 2\,NO_3^- \text{ (aq)}$$

Since 1 mol of $Ba(NO_3)_2$ yields 1 mol of Ba^{2+} and 2 mol of NO_3^-, it follows that a 1.00 M solution of $Ba(NO_3)_2$ is 1.00 M in Ba^{2+} and 2.00 M in NO_3^-.

3. *Exercises*

a. Give the concentration in M of each ion in 0.0238 M $Al_2(SO_4)_3$.

Solution: First we write the equation for the dissociation of the solid into ions. In this case

$$Al_2(SO_4)_3 \text{ (s)} \rightarrow 2\,Al^{3+} \text{ (aq)} + 3\,SO_4^{2-} \text{ (aq)}$$

Thus 1 mole of $Al_2(SO_4)_3$ yields 2 moles of Al^{3+} and 3 moles of SO_4^{2-}. The following conversion factors can be used.

$$\frac{1 \text{ mol } Al_2(SO_4)_3}{2 \text{ mol } Al^{3+}} \quad \text{and} \quad \frac{3 \text{ mol } SO_4^{2-}}{1 \text{ mol } Al_2(SO_4)_3}$$

Using these conversion factors, we obtain

$$\frac{0.0238 \text{ mol } Al_2(SO_4)_3}{1 \text{ L}} \times \frac{2 \text{ mol } Al^{3+}}{1 \text{ mol } Al_2(SO_4)_3} = 0.0476 \text{ M } Al^{3+}$$

and

$$\frac{0.0238 \text{ mol } Al_2(SO_4)_3}{1 \text{ L}} \times \frac{3 \text{ mol } SO_4^{2-}}{1 \text{ mol } Al_2(SO_4)_3} = 0.0714 \text{ M } SO_4^{2-}$$

b. Give the concentration, in moles per liter, of each ion in 0.0304 M Na_3N and 0.128 M $ScCl_3$. **(E3)**

II. Precipitation Reactions

A. Solubilities of ionic compounds

1. Anions whose compounds are generally soluble:

   a. All _____ are soluble.

   b. Chlorides are soluble except: _____

   c. Sulfates are soluble except: _____

2. Anions whose compounds are generally insoluble:

   a. Hydroxides are insoluble except: _____

   b. Carbonates are insoluble except: _____

   c. Phosphates are insoluble except: _____

Memorize these rules. In order to do well in this chapter, you need to master these rules.

3. *Exercises*

   a. Using the solubility rules shown in Figure 4.3 of your text, predict what will happen when aqueous solutions of silver nitrate and sodium phosphate are mixed.

   Solution: An aqueous solution of silver nitrate, $AgNO_3$, contains Ag^+ and NO_3^- ions; a solution of sodium phosphate, Na_3PO_4, contains Na^+ and PO_4^{3-} ions. Since an ionic compound is made up of an anion and a cation, two different ionic solids could be formed. They are

$$Ag_3PO_4 \quad \text{and} \quad NaNO_3$$

   From Figure 4.4, we see that $NaNO_3$ is soluble while Ag_3PO_4 is not. Thus, when the solutions are mixed, Ag_3PO_4 will precipitate.

   b. Predict what will happen when the following pairs of aqueous solutions are mixed: **(E4)**

   (1) solutions of copper sulfate and barium chloride.

   (2) solutions of ammonium phosphate and sodium hydroxide.

B. Equations for precipitation reactions

   1. Nomenclature
   Give the definition of

   a. Precipitate: _____

   _____

   b. Precipitation reaction: _____

   _____

   c. Net ionic equation: _____

   _____

   d. Spectator ions: _____

   _____

   2. Rules for writing net ionic equations

   a. Split up the ionic compounds into their respective ions. These are ions in solution.

   b. Combine a cation with an anion from the ions in solution. There are two possible combinations. These are the possible products.

   c. Check the solubility rules to determine which of the possible products is not soluble. This is the precipitate. There may be one, two, or no precipitates.

d. Write the reaction by putting the precipitate (from part **c**) on the product side first, and then the ions that make it up on the reactant side.
   Note: You will not be able to follow these steps if you still have not memorized the charges of metals and polyatomic ions (Chapter 2). Do so now!

3. *Exercises*

   a. Write the net ionic equation for the reaction that occurs when aqueous solutions of potassium carbonate and nickel(II) nitrate are mixed.

   Solution: We follow the steps outlined above.

   (1) <u>Ions in solution</u>: K^+, CO_3^{2-} (from potassium carbonate); and Ni^{2+}, NO_3^- (from nickel(II) nitrate).

   (2) <u>Possible products</u>: KNO_3; $NiCO_3$

   (3) <u>Precipitate</u>: The solubility rules indicate that only $NiCO_3$ is insoluble.

   (4) <u>Equation</u>: We first write the precipitate on the product side

$$\rightarrow NiCO_3 \text{ (s)}$$

   Then, we complete the reaction by writing the ions that make up the precipitate on the reactant side. Their subscripts are now coefficients.

$$Ni^{2+} \text{ (aq)} + CO_3^{2-} \text{ (aq)} \rightarrow NiCO_3 \text{ (s)}$$

   Remember: Always balance your net ionic equations.

   b. Write the net ionic equation for the precipitation reaction that occurs when aqueous solutions of ammonium phosphate and zinc nitrate are mixed. **(E5)**

C. Stoichiometry

   1. General notes on solving problems in solution stoichiometry

      a. Distinguish between the moles of reagent and the moles of reacting species. The reagent is the chemical you obtain from a bottle. The reacting species is most often an ion that is part of the reagent. For example, if 10.0 mL of 0.100 M $Ba(OH)_2$ are used in a reaction, then the data given allows you to calculate that 1.00×10^{-3} moles of $Ba(OH)_2$ are used. The reacting species, however is *not* $Ba(OH)_2$, but either Ba^{2+} or OH^- depending on the reaction. The number of moles of Ba^{2+} is equal to the number of moles of $Ba(OH)_2$ because there is one mole of Ba^{2+} in one mole of $Ba(OH)_2$. However, the number of moles of OH^- is not equal to the number of moles of $Ba(OH)_2$. There are 2 moles of OH^- for every mole of $Ba(OH)_2$, so in this case there are 2.00×10^{-3} moles of OH^- for the reaction.

      b. If you are asked to determine the concentration of ions after a reaction, you have to divide the number of moles of ions by the *total volume*. Total volume is the sum of the volumes of the solutions mixed together. We assume that for dilute aqueous solutions, volumes are additive.

c. The general procedure that should be followed is:

   (1) Write a balanced net ionic equation for the reaction.

   (2) Calculate the number of moles of reagent and relate it to the number of moles of reacting species.

   (3) Relate the number of moles of one reacting species to the number of moles of the other reacting species using the balanced net ionic equation.

   (4) Calculate the number of moles of product and convert it to the unit asked for.

2. Calculating the amount of product formed

  a. In this type of problem, complete information is usually given about both reactants, one of which is in excess. To determine what quantity or concentration of product is obtained, you must determine the limiting reactant. Do this by calculating the number of moles of each reacting species, and then calculating the moles of product assuming each reactant to be the limiting one.

  b. *Exercises*

   (1) When 200.0 mL of a 0.600 M solution of iron(III) chloride are mixed with 150.0 mL of a 0.100 M solution of barium hydroxide, a precipitate forms. How many grams of precipitate are formed?

   Solution: Using the steps described earlier, we solve the problem.

   — A precipitation reaction is involved.

    Ions in solution: Fe^{3+}; Cl^-; Ba^{2+}; OH^-

    Possible precipitates: $Fe(OH)_3$, $BaCl_2$

    Precipitate formed: $Fe(OH)_3$

    Net ionic equation: Fe^{3+} (aq) + $3\,OH^-$ (aq) $\rightarrow$ $Fe(OH)_3$ (s)

   — *moles* Fe^{3+}: Since Fe^{3+} comes from $FeCl_3$, we start with the information given about $FeCl_3$.

$$0.2000\text{ L} \times \frac{0.600\text{ mol FeCl}_3}{1\text{ L}} \times \frac{1\text{ mol Fe}^{3+}}{1\text{ mol FeCl}_3} = 0.120\text{ mol Fe}^{3+}$$

moles OH^-: The hydroxide ion is contributed by $Ba(OH)_2$, so we use the information given about it.

$$0.1500\text{ L} \times \frac{0.100\text{ mol Ba(OH)}_2}{1\text{ L}} \times \frac{2\text{ mol OH}^-}{1\text{ mol Ba(OH)}_2}$$

$$= 0.0300\text{ mol OH}^-$$

   — We now calculate the moles of $Fe(OH)_3$ obtained. Since we are mixing given amounts of both reactants, we must first find the limiting reactant.

   If all the Fe^{3+} is consumed:

$$0.120\text{ mol Fe}^{3+} \times \frac{1\text{ mol Fe(OH)}_3}{1\text{ mol Fe}^{3+}} = 0.120\text{ mol Fe(OH)}_3$$

If all the OH^- is consumed:

$$0.0300 \text{ mol } OH^- \times \frac{1 \text{ mol } Fe(OH)_3}{3 \text{ mol } OH^-} = 0.0100 \text{ mol } Fe(OH)_3$$

Since less $Fe(OH)_3$ is made using up OH^-, it is the limiting reactant, and the amount of $Fe(OH)_3$ produced is 0.0100 mol.

— We now calculate the mass of product formed.

$$0.0100 \text{ mol } Fe(OH)_3 \times \frac{106.9 \text{ g } Fe(OH)_3}{1 \text{ mol } Fe(OH)_3} = 1.07 \text{ g } Fe(OH)_3$$

(2) When 300.0 mL of a 0.500 M solution of calcium nitrate is combined with 200.0 mL of a 0.500 M solution of sodium carbonate, a precipitate forms. How many grams of precipitate are obtained? **(E6)**

3. Calculating how much of one reactant is required
   a. In this type of calculation, it is not necessary to determine the limiting reactant. Proceed using the steps described above, omitting the limiting reactant step.
   b. It is essential that you write out the net ionic equation before you start your calculations.
   c. *Exercises*
      (1) A solution of barium chloride is added to a copper(II) sulfate solution. A precipitate forms. What volume of 0.750 M copper(II) sulfate is required to react completely with 25.0 mL of 0.800 M barium chloride solution?
      Solution: The net ionic equation for this reaction is

$$Ba^{2+} \text{ (aq)} + SO_4^{2-} \text{ (aq)} \rightarrow BaSO_4 \text{ (s)}$$

We now follow the steps outlined earlier:

— We have complete information (volume and molarity) on barium chloride, so we calculate the number of moles.

$$0.0250 \text{ L} \times \frac{0.800 \text{ mol BaCl}_2}{1 \text{ L}} = 0.0200 \text{ mol BaCl}_2$$

— The reacting species, Ba^{2+}, comes from $BaCl_2$.

$$0.0200 \text{ mol BaCl}_2 \times \frac{1 \text{ mol Ba}^{2+}}{1 \text{ mol BaCl}_2} = 0.0200 \text{ mol Ba}^{2+}$$

— The other reacting species is SO_4^{2-}. According to the net ionic equation we can use the conversion factor

$$\frac{1 \text{ mol Ba}^{2+}}{1 \text{ mol SO}_4^{2-}}$$

Thus
$$0.0200 \text{ mol Ba}^{2+} = 0.0200 \text{ mol SO}_4^{2-}$$

— The reacting species, SO_4^{2-}, comes from $CuSO_4$. Hence

$$0.0200 \text{ mol SO}_4^{2-} \times \frac{1 \text{ mol CuSO}_4}{1 \text{ mol SO}_4^{2-}} = 0.0200 \text{ mol CuSO}_4$$

— We are asked for the volume of $CuSO_4$. We obtain this by using the molarity, 0.750 mol/L, as a conversion factor.

$$0.0200 \text{ mol CuSO}_4 \times \frac{1 \text{ L}}{0.750 \text{ mol CuSO}_4} = 0.0267 \text{ L}$$

We thus need 26.7 mL of 0.750 M $CuSO_4$ to react completely with 25.0 mL of 0.800 M $BaCl_2$.

(2) When 35.25 mL of a 0.125 M magnesium chloride solution react completely with 54.80 mL of potassium hydroxide solution, a precipitate forms. What is the molarity of the potassium hydroxide solution? **(E7)**

III. Acid-Base Reactions

A. Working definition of acid and base

1. Acid – a species that supplies H^+ ions to water
2. Base – a species that supplies OH^- ions to water

B. Types of acids

1. Strong acids

   a. Definition:

   A strong acid is one that completely ionizes in water forming an H^+ ion and an anion.

   b. *Example:*

   HNO_3 in water ionizes completely to form H^+ and NO_3^-. The reaction is

   $$HNO_3 \text{ (aq)} \rightarrow H^+ \text{ (aq)} + NO_3^- \text{ (aq)}$$

   In a solution prepared by adding 0.01 moles of HNO_3 to water there are 0.01 moles of H^+ and 0.01 moles of NO_3^- ions.

   c. What are the strong acids? (Name and formula)

   (1) _____

   (2) _____

   (3) _____

   (4) _____

   (5) _____

   (6) _____

   Memorize these six strong acids. You will have a lot of trouble with this chapter and subsequent chapters if you do not know them.

   Note that for all the strong acids, except H_2SO_4, one mole of acid produces one mole of H^+ ions. Sulfuric acid, in its reactions with bases, effectively produces two moles of H^+ ions per mole of H_2SO_4.

2. Weak acids

   a. Definition:

   A weak acid is one that only partially ionizes in water to give H^+ and the anion.

   b. *Example*

   HF partially ionizes in water to give H^+ and F^-. It is a weak acid. Its ionization equation is

   $$HF \text{ (aq)} \rightleftharpoons H^+ \text{ (aq)} + F^- \text{ (aq)}$$

   c. What are the weak acids?

   There are thousands of weak acids. The easiest way to determine whether a compound is a weak acid is to see whether ionization of the species yields an H^+ ion. If it does and it is not one of the six strong acids, then the acid must be weak. Any molecule that starts with an H is a prime candidate for a weak acid.

C. Types of bases
 1. Strong bases
   a. Definition:

   _____

   b. Example of the ionization of a strong base in water

   _____

   c. What are they?

   _____

   Know what the strong bases are. Note that for the hydroxides of Group 2
   metals, 2 OH$^-$ are produced for every mole of base that ionizes.
 2. Weak bases
   a. Weak bases do not furnish OH$^-$ ions by ionization. They react with water,
   and the OH$^-$ comes from the water. The remaining H$^+$ ion from the water
   attaches itself to the weak base. You may think of it as a two-step reaction:

   $$H_2O \rightleftharpoons H^+ (aq) + OH^- (aq)$$
   $$\text{weak base} + H^+ (aq) \rightleftharpoons (\text{weak base})H^+ (aq)$$

   $$\text{weak base} + H_2O \rightleftharpoons OH^- (aq) + (\text{weak base})H^+ (aq)$$

   b. What are the weak bases?
   In this chapter, the weak bases we will consider are ammonia, NH$_3$, and
   amines.
D. Writing equations for acid-base reactions
 1. Reacting species
   As we have seen, the species in aqueous solution depend on the type of acid or
   base (strong or weak). The reacting species are the species in water solution.
   They are written on the left side of an acid-base reaction. They are

   a. strong acid: _____

   b. strong base: _____

   c. weak acid: _____

   d. weak base: _____
 2. Net ionic equations
   a. Strong acid + strong base

   _____

   b. Strong acid + weak base

   _____

   c. Weak acid + strong base

   _____

3. *Exercises*

   a. Write the net ionic equation for the reaction between $H_3C_6H_5O_7$, citric acid, and sodium hydroxide.

   Solution: Citric acid is not one of the strong acids, thus it must be a weak acid. The reacting species is $H_3C_6H_5O_7$. Sodium hydroxide is a strong base. The reacting species is OH^-. The equation will be of the form

   $$HB\ (aq) + OH^-\ (aq)\ \rightarrow\ B^-\ (aq) + H_2O$$

   In this case, HB is $H_3C_6H_5O_7$ and B^- is $H_2C_6H_5O_7^-$. It is $H_3C_6H_5O_7$ minus an H. The net ionic equation is

   $$H_3C_6H_5O_7\ (aq) + OH^-\ (aq)\ \rightarrow\ H_2C_6H_5O_7^-\ (aq) + H_2O$$

   b. Write net ionic equations for the reactions between aqueous solutions of the following acids and bases.

   (1) phenylaniline, $C_6H_5NH_2$, and hydriodic acid **(E8)**

   (2) potassium hydroxide and perchloric acid **(E9)**

   (3) barium hydroxide and ascorbic acid, $HC_6H_7O_6$ **(E10)**

E. Acid-Base Titrations

   1. Definitions

   a. titration: _____

   b. standard solution: _____

   c. equivalence point: _____

   d. indicator: _____

2. Stoichiometry
   a. Review the general notes and the worked out exercises for the stoichiometry of precipitation reactions.
   b. *Exercises*
      (1) What is the molarity of an aqueous solution of $Ba(OH)_2$ if 45.00 mL are required to react with 12.35 mL of 0.1500 M HCl in order to reach the equivalence point? **(E11)**

      (2) A sample is known to contain only potassium hydroxide and an inert, nonreactive substance. If 29.80 mL of a 0.2513 M solution of perchloric acid reacts completely with the KOH in the sample, how many grams of KOH reacted? **(E12)**

      (3) Citric acid $(C_6H_8O_7)$ contains one mole of H^+/mole of citric acid. A sample containing citric acid has a mass of 1.286 g. The sample is dissolved in 100.0 mL of water. The solution is titrated with 0.0150 M NaOH. If 14.93 mL of the base are required to neutralize the acid, then what is the mass percent of citric acid in the sample? **(E13)**

(4) A sample of solid strontium hydroxide is mixed with water at 30°C and allowed to stand. A 100.0-mL sample of the solution is titrated with 59.4 mL of a 0.0400 M solution of hydrobromic acid. What is the concentration of the strontium hydroxide solution? **(E14)**

IV. Oxidation-Reduction Reactions
A. Some facts about redox reactions
1. Oxidation and reduction occur in the same reaction.

2. In a redox reaction the number of electrons (e⁻) lost must equal the number of electrons gained.

3. a. In a reduction reaction there is a(n) _____ of electrons.

   b. The electrons appear on the _____ side of the equation.

4. a. In an oxidation reaction there is a(n) _____ of electrons.

   b. The electrons appear on the _____ side of the equation.

B. Oxidation number
1. Oxidation number refers to the charge of a monatomic ion. In a molecule or polyatomic ion, each element gets a "pseudo-charge", arbitrarily obtained by assigning bonding electrons to the atom with the greater attraction for the electrons.

2. Rules for assigning oxidation number

   a. The oxidation number of an element in an elementary substance is _____.

   b. The oxidation number of an element in a monatomic ion is equal to

   _____ .

   c. Group 1 metals always have an oxidation number of _____ in their compounds.

   d. Group 2 metals always have an oxidation number of _____ in their compounds.

   e. The sum of the oxidation numbers of all the atoms in a neutral species is

   _____ .

f. The sum of the oxidation numbers of all the atoms in an ion is

_____ .

g. The oxidation number of hydrogen in compounds is almost always _____ . When it is combined with a metal only, then its oxidation number is -1.

h. The oxidation number of oxgen in compounds is almost always _____. When it is combined with a metal in Groups 1 and 2, determine its oxidation number using rules **(c)–(f)**.

Example: In Na_2O, oxygen has an oxidation number of -2. Since sodium is always $+1$ and there are two sodium ions, the cationic charge is $+2$. Hence, the anionic charge (the oxidation number of the oxygen atom) must be -2.

In Na_2O_2, sodium, having always an oxidation number of $+1$, makes the cationic charge $+2$. The total anionic charge is -2. There are two atoms of oxygen in the anion, so each atom is assigned an oxidation number of -1.

Note: Memorize these rules. Become adept at assigning oxidation numbers.

3. Exercises

    _a._ What is the oxidation number of each atom in SO_3^{2-}?

    Solution: Here, oxygen has oxidation number -2 since it is not combined with a Group 1 or Group 2 metal. Thus, if we assign x as the oxidation number for sulfur, and recognize that the charge for the whole polyatomic ion is -2 (see superscript), we get the equation

$$x + 3(-2) = -2$$
$$x = 4$$

The oxidation number for S in SO_3^{2-} is $+4$.

    _b._ What are the oxidation numbers for all atoms in NO_2 and NH_4^+? **(E15)**

4. Defining oxidation-reduction on the basis of oxidation numbers

    a. Oxidation can be defined as a(n) _____ in oxidation number.

    b. Reduction can be defined as a(n) _____ in oxidation number.

    c. The reducing agent is the substance being oxidized.

    d. The oxidizing agent is the substance being reduced.

e. Exercises

    (1) For the reaction

$$2\,Cl_2\,(aq) + 2\,H_2O\,(g) \rightarrow 4\,HCl\,(g) + O_2\,(g)$$

identify the oxidizing and reducing agents.

Solution: To do this, we must know which of the species is oxidized and which is reduced. Thus, we determine the oxidation number of each of the atoms in all the compounds. The oxidation numbers are:

$$\text{reactants}: Cl = \;\;\;0; \quad H = +1; \quad O = -2$$
$$\text{products}: Cl = -1; \quad H = +1; \quad O = \;\;\;0$$

We see that chlorine goes from an oxidation number of 0 to −1. Since that is reduction (because there is a decrease in oxidation number), chlorine is the species reduced and Cl_2 is the oxidizing agent.
We see that hydrogen's oxidation number stays the same.
We also note that oxygen goes from an oxidation number of −2 to 0. That is an increase in oxidation number, and oxygen is said to be oxidized. Hence H_2O is the reducing agent.

    (2) In the unbalanced redox equation

$$MnO_4^-\,(aq) + NO_2^-\,(aq) + H^+\,(aq) \rightarrow Mn^{2+}\,(aq) + NO_3^-\,(aq) + H_2O$$

identify the oxidizing and reducing agents. **(E16)**

C. Balancing half-reactions (oxidation and reduction)
The following steps should be followed in balancing oxidation and reduction half-reactions. Follow them systematically every time you have to balance a half-reaction, so that the process becomes a habit.
1. Write an oxidation half-reaction by picking out the ions and/or compounds that contain the element oxidized (i.e., the one that increases its oxidation number). Do the same for the reduction half-equation, only this time pick out the ions and/or compounds that contain the element reduced (i.e., the one that decreases its oxidation number).
2. Balance the atoms of the element being oxidized or reduced.
3. Add electrons to reflect the change in oxidation number.
    a. Put the electrons on the left side when reduction occurs.
    b. Put the electrons on the right side when oxidation occurs.

c. The number of electrons should reflect the *total* change in oxidation number. *Example:* For the reduction half-reaction

$$Cl_2 (g) \rightarrow 2\,Cl^- (aq)$$

the change in oxidation number is from 0 (Cl_2) to -1 (Cl^-). However, since there are two Cl^- ions, each with an oxidation number of -1, the change in oxidation number is 2. Thus the half-reaction should be written

$$Cl_2 (g) + 2\,e^- \rightarrow 2\,Cl^- (aq)$$

4. Balance charge.
   a. Add H^+ ions to the side with the smaller total charge when the reaction is done in acidic medium.
   b. Add OH^- ions to the side with the larger total charge when the reaction is done in basic medium.
5. Balance atoms.
   a. The element that is either oxidized or reduced should already be balanced (Step **1**).
   b. Add H_2O to balance hydrogen.
   c. Check to see that oxygen is balanced. It should be.

All this will make more sense when you see it actually used and when you use it yourself.

6. *Exercises*
   a. Consider the following redox reaction in acidic medium.

$$Cr_2O_7^{2-} (aq) + NO_2^- (aq) \rightarrow Cr^{3+} (aq) + NO_3^- (aq)$$

Write balanced oxidation and reduction half-reactions.
Solution: Let's start with the reduction half-reaction.
Step (1) Chromium is the element reduced since it has an oxidation number of +6 as reactant and +3 as product. The half-reaction is

$$Cr_2O_7^{2-} \rightarrow Cr^{3+}$$

Note that $Cr_2O_7^{2-}$ is the reactant. It is the species that has the Cr atom. Do *not* write

$$Cr^{6+} \rightarrow Cr^{3+}$$

as your half-reaction.
Step (2) Since there are 2 Cr atoms on the left and only one on the right, we write

$$Cr_2O_7^{2-} \rightarrow 2\,Cr^{3+}$$

Step (3) The half-reaction is a reduction so electrons (e^-) are put on the left side. The oxidation number goes from $2(+6) = 12$ (since there are two chromium on the left each with an oxidation number of +6) to 2(+3) = 6 (again since there are 2 chromium on the right, each with an oxidation number of +3). The change in oxidation number is from +12 to +6 which is $6e^-$.

$$Cr_2O_7^{2-} + 6e^- \rightarrow 2\,Cr^{3+}$$

Step (4) We now balance charge. The charge on the left is −8 [i.e.,−2 + (−6)]. The charge on the right is +6 [i.e., 2 x (+3)]. The smaller charge is on the left (i.e., −8 < +6) so we add H^+ to the left side. To figure out how many H^+ you have to add, you can use the algebraic equation

$$-8 + x = +6$$

where −8 is the charge on the left, x is the number of H^+ to be added, and +6 is the charge on the right. We see that x = 14. (If you get a negative number of x, you have chosen the wrong side for x.) Thus, we need 14 H^+, and we write

$$Cr_2O_7^{2-} + 6e^- + 14\,H^+ \rightarrow 2\,Cr^{3+}$$

The charge is now balanced. On the left we have a total charge of +6 [i.e., −6 + (−2) + 14], on the right we have +6 [i.e., 2(+3)].
Step (5) We balance atoms.
(1) Cr was balanced in Step (1).
(2) For hydrogen, we have 14 H on the left and no H on the right. We need to add H_2O, and since we need 14 H on the right, we add 7 H_2O on the right.

$$Cr_2O_7^{2-} + 6e^- + 14\,H^+ \rightarrow 2\,Cr^{3+} + 7\,H_2O$$

(3) We see that oxygen has also been balanced by adding water.
The balanced reduction half-reaction is

$$Cr_2O_7^{2-}\,(aq) + 6e^- + 14\,H^+\,(aq) \rightarrow 2\,Cr^{3+}\,(aq) + 7\,H_2O$$

We now write the oxidation half-reaction.
Step (1) Nitrogen is the element oxidized from +3 to +5. NO_2^- has the N atom and is written on the left. NO_3^- also has the N atom, and is written on the right.

$$NO_2^- \rightarrow NO_3^- -$$

Step (2) There is one N on the left and one on the right, so the atom that changes oxidation number is balanced.
Step (3) The oxidation number changes from 1(3) = 3 to 1(5) = 5. The total change in oxidation number is 2. Since this is an oxidation half-reaction, we write $2e^-$ on the right.

$$NO_2^- \rightarrow NO_3^- + 2\,e^-$$

Step (4) We balance charge. The total charge on the left is -1. The total charge on the right is -3. (Do not confuse charge with oxidation number!) The smaller charge is on the right ($-3 < -1$), so we add H^+ (x) on the right.

$$-1 = -3 + x$$
$$x = 2$$

Thus we add 2 H^+ on the right side.

$$NO_2^- \ \rightarrow \ NO_3^- + 2\,e^- + 2\,H^+$$

Step (5) We balance hydrogen atoms. There are 2 H on the right but none on the left. We balance that by adding an H_2O on the left.

$$NO_2^- + H_2O \ \rightarrow \ NO_3^- + 2\,e^- + 2\,H^+$$

We see that the oxygen atoms have also been balanced. The oxidation half-reaction is

$$NO_2^-\,(aq) + H_2O \ \rightarrow \ NO_3^-\,(aq) + 2\,H^+\,(aq) + 2\,e^-$$

b. For the reaction

$$PbO\,(s) + NH_3\,(g) \ \rightarrow \ Pb\,(s) + N_2\,(g) + H_2O\,(g)$$

Write a balanced reduction half-reaction and oxidation half-reaction in basic medium. **(E17)**

D. Balancing redox reactions

    1. Steps to balance overall redox reactions

        a. *Step 1:* Split the equation and write a balanced oxidation half-reaction and a balanced reduction half-reaction.
This process was just explained.

        b. *Step 2:* Multiply the two half-reactions by coefficients so that the two half-reactions have the same number of electrons on opposite sides.

        c. *Step 3:* Add the two equations making sure that

          (1) electrons cancel out.

          (2) H^+, OH^-, and/or H_2O appear on only one side of the overall equation.

    2. *Exercises*

        a. Write a balanced net ionic equation for the following redox reaction in acidic medium.

$$Cr_2O_7^{2-} \text{ (aq)} + NO_2^- \text{ (aq)} \rightarrow Cr^{3+} \text{ (aq)} + NO_3^- \text{ (aq)}$$

Solution:

Step (1): The balanced half-reactions (explained earlier) are

$$Cr_2O_7^{2-} \text{ (aq)} + 6\,e^- + 14\,H^+ \text{ (aq)} \rightarrow 2\,Cr^{3+} \text{ (aq)} + 7\,H_2O$$

$$NO_2^- \text{ (aq)} + H_2O \rightarrow NO_3^- \text{ (aq)} + 2\,H^+ \text{ (aq)} + 2\,e^-$$

Step (2): Since both half-reactions must have the same number of electrons, we multiply the oxidation half-reaction (NO_2^- (aq) $\rightarrow$ NO_3^- (aq)) by 3. We leave the reduction half-reaction alone. Doing that we get

$$Cr_2O_7^{2-} \text{ (aq)} + 6\,e^- + 14\,H^+ \text{ (aq)} \rightarrow 2\,Cr^{3+} \text{ (aq)} + 7\,H_2O$$

$$3\,NO_2^- \text{ (aq)} + 3\,H_2O \rightarrow 3\,NO_3^- \text{ (aq)} + 6\,H^+ \text{ (aq)} + 6\,e^-$$

Step (3): We add both half-reactions, cancelling the 6 electrons appearing on both sides.

$$Cr_2O_7^{2-} \text{ (aq)} + 3\,NO_2^- \text{ (aq)} + 14\,H^+ \text{ (aq)} + 3\,H_2O \rightarrow$$
$$2\,Cr^{3+} \text{ (aq)} + 7\,H_2O + 3\,NO_3^- \text{ (aq)} + 6\,H^+ \text{ (aq)}$$

Note that

(1) 14 H^+ appear on the left and 6 H^+ appear on the right. Net H^+ is 8 H^+ on the left. (Subtract 6 H^+ from both sides.)

(2) 3 H_2O appear on the left and 7 H_2O appear on the right. Net H_2O is 4 H_2O on the right. (Subtract 3 H_2O from both sides.)

The balanced net ionic redox reaction is

$$Cr_2O_7^{2-} \text{ (aq)} + 3\,NO_2^- \text{ (aq)} + 8\,H^+ \text{ (aq)} \rightarrow$$
$$2\,Cr^{3+} \text{ (aq)} + 4\,H_2O + 3\,NO_3^- \text{ (aq)}$$

b. Write a balanced net ionic equation for the following redox reaction

$$S_2O_3^{2-} \text{ (aq)} + Cl_2 \text{ (g)} \rightarrow SO_4^{2-} \text{ (aq)} + Cl^- \text{ (aq)}$$

in both acidic and basic medium. **(E18)**

E. Stoichiometry

1. Stoichiometric calculations for redox reactions are carried out in much the same way as those for acid-base or precipitation reactions.

2. *Exercises*

   a. When aluminum is added to a strong acid, the metal is oxidized to aluminum(III) ions by H^+ and hydrogen gas is produced. If 10.0 g of aluminum is added to 50.0 mL of 0.450 M HCl, what is the concentration of Al^{3+} in moles/liter after the reaction is complete? Assume no volume change. **(E19)**

b. All the iron in an ore with mass 1.500 g is converted to Fe^{2+}. The Fe^{2+} ions are made to react with a $K_2Cr_2O_7$ solution. The reaction required 35.25 mL of a 0.05000 M solution of $K_2Cr_2O_7$. The reaction is

$$6\,Fe^{2+}\,(aq) + Cr_2O_7^{2-}\,(aq) + 14\,H^+\,(aq) \rightarrow$$
$$6\,Fe^{3+}\,(aq) + 2\,Cr^{3+}\,(aq) + 7\,H_2O$$

Calculate the percent of Fe in the ore. **(E20)**

SELF-TEST
A. Multiple choice:

1. Which of the following substances would precipitate from a mixture of barium, silver, chloride, sulfate, and nitrate ions in aqueous solution?

     (1) $BaSO_4$ (2) $BaCl_2$ (3) $Ba(NO_3)_2$ (4) Ag_2SO_4 (5) AgCl

a. (2)   **b.** (1),(5)   **c.** (1),(2),(3)

d. (1),(2),(4)   **e.** (1),(2),(5)   **f.** (1),(2),(4),(5)

2. The acid-base reaction between aqueous solutions of HNO_2 and KOH

     (1) will produce NO_2 (g) and NO (g).
     (2) is a precipitation reaction.
     (3) is a weak acid – strong base reaction.
     (4) will produce only water.
     (5) will produce NO_2^- (aq).

The correct statements are
a. (1),(2)   **b.** (1),(3)   **c.** (2),(5)   **d.** (3),(5)   **e.** (3),(4)

3. In which species does sulfur have the same oxidation number as chlorine in ClO_2^-?
a. H_2S **b.** S_8 **c.** SO_3^{2-} **d.** SO_4^{2-} **e.** none of thes

4. In an acid-base reaction, the reacting species for $Ba(OH)_2$ is
a. H^+ **b.** OH^- **c.** $(OH)_2^-$ **d.** BaO   **e.** Ba^{2+}

5. When NaOH was added to one liter of a solution containing 0.10 mol of each of two metal ions, a precipitate formed with one ion but the other stayed in solution. The solution could have contained
   a. Mg^{2+} and Ba^{2+} b. K^+ and Ba^{2+} c. Mg^{2+} and Cr^{3+}
   d. Cr^{3+} and Fe^{3+} e. none of these

6. The balanced equation for the reaction between NO_3^- and Br^- in acid is:

$$2\,NO_3^-\,(aq) + 4\,H^+\,(aq) + 2\,Br^-\,(aq) \rightarrow 2\,NO_2\,(g) + Br_2\,(\ell) + 2\,H_2O$$

When this equation is balanced in basic solution, the number of H_2O molecules is
   a. 2 on the left   b. 2 on the right
   c. 4 on the left   d. 4 on the right
   e. zero on both sides of the equation

7. Consider the unbalanced equation for the following reaction in acid medium:

$$Se\,(s) + NO_3^-\,(aq) \rightarrow SeO_2\,(s) + NO\,(g)$$

In the balanced, net ionic equation, the smallest whole number coefficients for Se and NO_3^- respectively are
   a. 3,4   b. 4,3   c. 2,4   d. 4,2   e. 2,3

8. How many of the following anions are derived from strong acids?
   Cl^- $C_2H_3O_2^-$ PO_4^{3-} NO_3^- CO_3^{2-}
   a. 1   b. 2   c. 3   d. 4   e. 5

9. For the reaction

$$Cu\,(s) + 4\,H^+\,(aq) + 2\,NO_3^-\,(aq) \rightarrow 2\,NO_2\,(g) + 2\,H_2O + Cu^{2+}\,(aq)$$

the oxidizing agent and the reducing agent are, in that order:
   a. Cu, NO_3^- b. Cu, H^+ c. H^+
   d. H^+, Cu e. NO_3^-, Cu

10. The equation for a redox reaction in basic solution
    a. always has hydroxide ions as reactant.
    b. may have hydroxide ions as either reactant or product.
    c. may contain hydrogen ions.
    d. always has water as a product.
    e. None of the above is true.

B. Writing Equations:

Write balanced net ionic equations using smallest whole number coefficients for the following reactions between 0.1 M aqueous solutions of the following:

1. lead nitrate and sodium chloride

_____

2. perchloric acid and barium hydroxide

_____

3. hydriodic acid and ammonia

_____

4. acetic acid and potassium hydroxide

_____

5. sulfuric acid and barium chloride

_____

6. nickel(II) nitrate and hydrochloric acid

_____

C. Problems:

1. Consider mixing solutions of silver nitrate and calcium chloride.
   a. Write a net ionic equation for the formation of the precipitate, if any.

   b. If 200.0 mL of 0.300 M silver nitrate are mixed with 350.0 mL of 0.500 M calcium chloride, how many grams of precipitate are formed? How many moles of each ion are present after precipitation?

   c. What volume of 0.250 M calcium chloride would be required to precipitate silver chloride if it is mixed with 600.0 mL of 0.650 M silver nitrate?

2. Consider two solutions. One solution is 0.1115 M $Ca(OH)_2$. The other is 0.1050 M $HClO_4$.

   a. Write a balanced equation for the reaction between the two solutions.

   b. How many mL of $Ca(OH)_2$ will be required to neutralize 25.00 mL of the $HClO_4$?

   c. If a student starts to titrate 31.39 mL of the $Ca(OH)_2$ solution with $HClO_4$ and stops the titration after only 23.81 mL of $HClO_4$ have been added, then
   (1) How many moles of H^+ have been added?

   (2) How many moles of OH^- are left unreacted?

3. Balance the following equations:

    a. $S_2O_3^{2-}$ (aq) + I_3^- (aq) $\rightarrow$ $S_4O_6^{2-}$ (aq) + I^- (aq)

    b. Mn^{2+} (aq) + $HBiO_3$ (s) $\rightarrow$ MnO_4^- (aq) + BiO^+ (aq)   (acid)

    c. $Cr_2O_7^{2-}$ (aq) + H_3AsO_3 (aq) $\rightarrow$ H_3AsO_4 (aq) + Cr^{3+} (aq)   (acid)

    d. ClO^- (aq) + CrO_2^- (aq) $\rightarrow$ Cl^- (aq) + CrO_4^{2-} (aq)   (basic)

    e. OBr^- (aq) + HPO_3^{2-} (aq) $\rightarrow$ Br^- (aq) + PO_4^{3-} (aq)   (basic)

4. The salt $NaBrO_3$ oxidizes Sn^{2+} to $SnCl_6^{2-}$ in the presence of hydrochloric acid according to the equation

$$3\,Sn^{2+}\,(aq) + 18\,Cl^-\,(aq) + BrO_3^-\,(aq) + 6\,H^+\,(aq) \rightarrow$$
$$3\,SnCl_6^{2-}\,(aq) + Br^-\,(aq) + 3\,H_2O$$

A sample weighing 2.000 g is dissolved in acid and all the tin present is converted to Sn^{2+}. For the reaction to go to completion, 32.50 mL of a 0.07500 M $KBrO_3$ solution are required. Find the percent of tin in the sample.

ANSWERS
Exercises:
(E1) 102 mL   **(E2)** 1.402 M

(E3) 0.0912 M Na^+, 0.0304 M N^{3-}, 0.128 M Sc^{3+}, 0.384 M Cl^-

(E4) (1) A precipitate of barium sulfate forms.   (2) No precipitate forms.

(E5) $3\,Zn^{2+}\,(aq) + 2\,PO_4^{3-}\,(aq) \rightarrow Zn_3(PO_4)_2\,(s)$

(E6) 10.0 g   **(E7)** 0.161 M

(E8) $C_6H_5NH_2\,(aq) + H^+\,(aq) \rightarrow C_6H_5NH_3^+\,(aq)$

(E9) $H^+\,(aq) + OH^-\,(aq) \rightarrow H_2O$

(E10) $HC_6H_7O_6\,(aq) + OH^-\,(aq) \rightarrow C_6H_7O_6^-\,(aq) + H_2O$

(E11) 0.02058 M   **(E12)** 0.4202 g   **(E13)** 3.34 %

(E14) 0.0119 M

(E15) In NO_2: N has oxidation number +4 and O has oxidation number -2.
In NH_4^+: N has oxidation number -3 and H has oxidation number +1.

(E16) oxidizing agent - MnO_4^-; reducing agent - NO_2^-

(E17) reduction half-reaction: $PbO\,(s) + 2\,e^- + H_2O \rightarrow Pb\,(s) + 2\,OH^-\,(aq)$
oxidation half-reaction: $2\,NH_3\,(g) + 6\,OH^-\,(aq) \rightarrow N_2\,(g) + 6\,e^- + 6\,H_2O$

(E18) acid: $4\,Cl_2\,(g) + S_2O_3^{2-}\,(aq) + 5\,H_2O \rightarrow 8\,Cl^-\,(aq) + 2\,SO_4^{2-}\,(aq) + 10\,H^+\,(aq)$
base: $4\,Cl_2\,(g) + S_2O_3^{2-}\,(aq) + 10\,OH^-\,(aq) \rightarrow 8\,Cl^-\,(aq) + 2\,SO_4^{2-}\,(aq) + 5\,H_2O$

(E19) 0.150 M   **(E20)** 39.37 %

Self-Test
A. Multiple Choice:

1. b	**2.** d	**3.** e	**4.** b	**5.** a
6. a	**7.** a	**8.** b	**9.** e	**10.** b

B. Writing Equations:

1. $Pb^{2+}(aq) + 2\,Cl^-(aq) \rightarrow PbCl_2(s)$

2. $H^+(aq) + OH^-(aq) \rightarrow H_2O$

3. $H^+(aq) + NH_3(aq) \rightarrow NH_4{}^+(aq)$

4. $HC_2H_3O_2(aq) + OH^-(aq) \rightarrow H_2O + C_2H_3O_2^-(aq)$

5. $Ba^{2+}(aq) + SO_4^{2-}(aq) \rightarrow BaSO_4(s)$

6. no reaction

C. Problems:

1. a. $Ag^+(aq) + Cl^-(aq) \rightarrow AgCl(s)$

   b. 8.60 g AgCl; 0.00 mol Ag^+; 0.290 mol Cl^-; 0.175 mol Ca^{2+}; 0.0600 mol NO_3^-

   c. 0.780 L

2. a. $H^+(aq) + OH^-(aq) \rightarrow H_2O$

   b. 11.77 mL

   c. (1) 2.500×10^{-3} mol H^+ (2) 4.500×10^{-3} mol OH^-

3. a. $2\,S_2O_3^{2-}(aq) + I_3^-(aq) \rightarrow S_4O_6^{2-}(aq) + 3\,I^-(aq)$

   b. $5\,HBiO_3(s) + 2\,Mn^{2+}(aq) \rightarrow 2\,MnO_4^-(aq) + 5\,BiO^+(aq) + 2\,H_2O + H^+(aq)$

   c. $Cr_2O_7^{2-}(aq) + 3\,H_3AsO_3(aq) + 8\,H^+(aq) \rightarrow$
$$2\,Cr^{3+}(aq) + 3\,H_3AsO_4(aq) + 4\,H_2O$$

   d. $3\,ClO^-(aq) + 2\,CrO_2^-(aq) + 2\,OH^-(aq) \rightarrow 3\,Cl^-(aq) + 2\,CrO_4^{2-}(aq) + H_2O$

   e. $OBr^-(aq) + HPO_3^{2-}(aq) + OH^-(aq) \rightarrow Br^-(aq) + PO_4^{3-}(aq) + H_2O$

4. 43.40 %

5
Gases

I. Measurements on Gases
  A. Volume
    1. Units
      a. liter (L)
      b. milliliter (mL)
      c. cubic centimeter (cm^3)
      d. cubic meter (m^3)
    2. Equivalences

  1 L = _____ mL.

  1 L = _____ cm^3.

  1 L = _____ m^3.

  B. Amount
    1. Unit/Symbol
      a. moles (mol)/n
      b. mass in grams/m
    2. Equivalence
      $n = \dfrac{m}{MM}$
  C. Temperature
    1. Unit
      a. Kelvin (K)
      b. Centigrade or Celsius (°C)
    2. Equivalence
      K = (°C) + 273.15
  D. Pressure
    1. Unit
      a. millimeters of mercury (mm Hg)
      b. atmosphere (atm)

    c. bar
  2. Equivalences
      1 atm = _____ mm Hg

      1 atm = _____ bar

II. Ideal Gas Law
  A. Mathematical Expression
    1. $PV = nRT$

      P = _____

      V = _____

      n = _____

      R = _____

      T = _____
    2. n is always in moles.
      T is always in Kelvin.
    3. R – gas constant
      If P is in atm and V is in L, then R is _____. This is the most
      commonly used R.
  B. Final and initial state problems
    1. Steps to follow in solving these problems:
      a. List all variables (P, V, n, T) in two columns, one for the initial state, the other
        for the final state.
      b. Write two gas equations, one for the initial state, the other for the final state.
        Give subscript 1 for the initial state and 2 for the final state for the variables
        that vary. Those that remain constant do not get any subscript.
      c. Rewrite both gas equations so that subscripted variables are on the left
        and nonsubscripted variables are on the right. Equate left sides of both
        equations to each other.
      d. Substitute into the single equation obtained in (c) the variables given for both
        initial and final state, and solve for the desired variable.
    2. Exercises
      a. A sample of gas occupies 355 mL at 15 °C and 755 mm Hg pressure. What
        temperature will the gas have at the same pressure if its volume increases
        to 453 mL?
        Solution:
        (1) List all variables (P, V, n, T) in two columns, one for the initial state, the
          other for the final state.

             P = 755 mm Hg   P = 755 mm Hg
             V = 355 mL   V = 453 mL
             T = 288 K   T = ?
             n = constant   n = constant

As you can see, volume and temperature change, whereas pressure and amount remain constant.

(2) Our two gas equations are therefore

$$PV_1 = nRT_1 \qquad PV_2 = nRT_2$$

We did not give the subscripts 1 and 2 to P and n because P and n remain constant.

(3) Putting all constants (no subscripts) on one side and all variables (with subscripts) on the other, we get

$$\frac{V_1}{T_1} = \frac{nR}{P} \qquad \frac{V_2}{T_2} = \frac{nR}{P}$$

resulting in the single equation

$$\frac{V_1}{T_1} = \frac{V_2}{T_2}$$

(4) Substituting, we get

$$\frac{355 \text{ mL}}{288 \text{ K}} = \frac{453 \text{ mL}}{T_2}$$

$$T_2 = \frac{288 \text{ K} \times 453 \text{ mL}}{355 \text{ mL}} = 368 \text{ K}$$

A word about units: For initial and final state problems, do not worry about making units compatible with those of R. You only have to make sure that the units of the initial state and those of the final state remain the same. Temperature, however, should always be in Kelvin.

b. A sample of gas at 30.0°C has a pressure of 1.23 bar and a volume of 0.0247 m³. If the volume of the gas is compressed to 0.00839 m³ at the same temperature, what is its pressure at this volume? **(E1)**

c. A sample of gas at 27.0°C has a volume of 2.08 L and a pressure of 750.0 mm Hg. If the gas is in a sealed container, what is its pressure in bar when the temperature (in °C) doubles? **(E2)**

C. One-state problems

  1. One-state or single-state problems involve R and the variables P, V, n, and T. In these problems one measurement for each of three variables is given and you are asked to solve for the fourth variable. The units for the R that you use will determine the units that you will have to use for the variables. Remember T must always be in Kelvin.

  2. Exercises:

  a. How many grams of oxygen gas in a 10.0-L container will exert a pressure of 712 mm Hg at a temperature of 25.0°C?

  Solution: Since R = 0.0821 L·atm/K·moles, V has to be in liters, P in atmospheres, T in Kelvin and n in moles. In this problem

  V = 10.0 L

  $P = 712 \text{ mm Hg} \times \dfrac{1 \text{ atm}}{760 \text{ mm Hg}} = 0.937 \text{ atm}$

  T = 25.0°C + 273.15 = 298.1 K

  n = unknown variable asked for in grams

  Substituting into the ideal gas equation, we get

$$PV = nRT$$

$$n = \frac{0.937 \text{ atm} \times 10.0 \text{ L}}{0.0821 \dfrac{\text{L} \cdot \text{atm}}{\text{mol} \cdot \text{K}} \times 298.1 \text{ K}} = 0.383 \text{ mol } O_2$$

Since the amount of gas has to be expressed in grams of oxygen, we use the conversion factor

$$32.0 \text{ g} = 1 \text{ mol}$$

Thus

$$0.383 \text{ mol} \times \frac{32.0 \text{ g}}{1 \text{ mol}} = 12.3 \text{ g } O_2$$

b. At what temperature will 26.42 g of sulfur dioxide in a 250.0-mL container exert a pressure of 0.842 atm? **(E3)**

D. Other calculations involving the Ideal Gas Law
   1. Molar mass calculations
      a. We can often determine the molar mass of a gas by measuring its mass at a given volume, temperature, and pressure. To do this, we simply write m/MM instead of n in the ideal gas equation. We therefore get

$$PV = \frac{m}{MM}RT \quad \text{instead of} \quad PV = nRT$$

where MM is the molar mass, and m is the mass in grams. Thus

$$MM = \frac{mRT}{PV}$$

If you remember to keep your units straight, this type of problem becomes a simple "plug-in".

      b. *Exercises*
         (1) When 4.93 g of carbon tetrachloride gas are in a 1.00-L container at 400.0 K, the gas exerts a pressure of 800.0 mm Hg. What is the molar mass of carbon tetrachloride?
         Solution: Using the equation MM = mRT/PV, we get

$$MM = \frac{4.93 \text{ g} \times 0.0821 \frac{\text{L} \cdot \text{atm}}{\text{mol} \cdot \text{K}} \times 400.0 \text{ K}}{1.00 \text{ L} \times 800.0 \text{ mm} \times \frac{1 \text{ atm}}{760 \text{ mm}}} = 154 \frac{\text{g}}{\text{mol}}$$

To check, we know that carbon tetrachloride is written as CCl_4. Thus, its molar mass must be 12.0 + 4(35.45) = 153.8. Remember that the compound with the smallest molar mass is hydrogen (2.02 g/mol). Therefore, if you arrive at a molar mass of less than 2.02 g/mol, you are doing something wrong!

(2) A gaseous compound with a mass of 3.216 g has a volume of 2236 mL at 27.0°C and 735 mm Hg. Calculate the approximate molar mass of the compound. **(E4)**

2. Density calculations
   a. The equation PV = nRT can also be used to calculate the density of a gas at given conditions of temperature and pressure.
   The variable to find is n.
   Since density is the mass of gas that occupies one liter, then V = 1.00 L.
   After finding n, change n to mass and you get mass/L.
   b. Given the density of a gas at conditions of temperature and pressure, one can also determine the molar mass of the gas.
   Take the mass portion of the density value and call that m.
   The volume is 1.00 L.
   You can then use the expression

$$PV = \frac{m}{MM} RT$$

   to find MM.
   c. *Exercises*
   (1) Calculate the density of ammonia at 0°C and 1 atm pressure.
   Solution: Here P = 1.00 atm, T = 273 K and V = 1.00 L. We find n using the ideal gas equation, PV = nRT:

$$n = \frac{1.000 \text{ atm x } 1.000 \text{ L}}{0.0821 \dfrac{\text{L} \cdot \text{atm}}{\text{mol} \cdot \text{K}} \text{ x } 273 \text{ K}} = 0.0446 \text{ mol}$$

   We now use the molar mass of NH_3 (17.03 g/mol) to determine how many grams are equivalent to 0.0446 mol.

$$0.0446 \text{ x } 17.03 \frac{\text{g}}{\text{mol}} = 0.760 \frac{\text{g}}{\text{L}}$$

Note that the densities of gases are expressed as g/L, whereas those of liquids and solids are expressed in g/mL.

(2) Calculate the molar mass of a compound with a density of 0.630 g/L at 25.0°C and 730.0 mm Hg pressure. **(E5)**

III. Stoichiometry of Reactions Involving Gases

A. Volumes at specified temperature and pressure measurements

Since gases are often measured in liters or milliliters instead of grams, you need to convert that volume into either grams or moles in order to solve stoichiometric problems involving gases.

Exercises

1. Consider the reaction

$$MnO_2 \text{ (s)} + 2\,Cl^- \text{ (aq)} + 4\,H^+ \text{ (aq)} \rightarrow Mn^{2+} \text{ (aq)} + Cl_2 \text{ (g)} + 2\,H_2O$$

How many grams of MnO_2 are required to produce 1.200 L of Cl_2 gas at 1.00 atm and 200°C?

Solution: Here you want to convert 1.200 L of Cl_2 (at 1.00 atm and 200°C) to grams of MnO_2. For this reaction, the equivalences are

$$1 \text{ mol } MnO_2 = 1 \text{ mol } Cl_2 = 86.94 \text{ g } MnO_2 = 70.90 \text{ g } Cl_2$$

Using the ideal gas law you can determine the number of moles of Cl_2.

$$n = \frac{1 \text{ atm} \times 1.200 \text{ L}}{0.0821\ \dfrac{L \cdot atm}{mol \cdot K} \times 473 \text{ K}} = 0.0309 \text{ mol } Cl_2$$

Now you can use a conversion factor (from the equivalences above) to change moles of Cl_2 to grams of MnO_2.

$$0.0309 \text{ mol } Cl_2 \times \frac{86.94 \text{ g } MnO_2}{1 \text{ mol } Cl_2} = 2.69 \text{ g } MnO_2$$

2. When sodium metal reacts with water, sodium ions, hydroxide ions, and hydrogen gas are produced.

a. Write a balanced equation for the reaction.

  b. How many liters of hydrogen gas will be collected at 25.00°C and 1.00 atm pressure if 2.15 g of sodium are used? **(E6)**

B. Volumes measured when P and T remain constant

Law of Combining Volumes: _____

_____

This law can be applied to the reaction

$$2\,H_2\ (g) + O_2\ (g)\ \rightarrow\ 2\,H_2O\ (g)$$

in the following way:

   2 mol H_2 = 1 mol O_2 = 2 mol H_2O = 4.04 g H_2 = 32.0 g O_2 = 36.0 g H_2O

or NOT *and*

$$2\,L\ H_2 = 1\,L\ O_2 = 2\,L\ H_2O$$

You may NOT write 36.0 g H_2O = 2 L H_2O. The volume equivalences may be written only if the volumes are measured *at the same temperature and pressure.* The temperature and pressure do not need to be specified, but the problem has to state that the conditions remain the same.

Exercises

1. Consider the reaction

$$2\ NO\ (g) + O_2\ (g)\ \rightarrow\ 2\,NO_2\ (g)$$

What volume of NO_2 will be produced by 1.38 L of O_2 when both volumes are measured at the same temperature and pressure?

Solution:

$$2 \text{ L NO} = 1 \text{ L O}_2 = 2 \text{ L NO}_2$$

$$1.38 \text{ L O}_2 \times \frac{2 \text{ L NO}_2}{1 \text{ L O}_2} = 2.76 \text{ L NO}_2$$

2. Consider the reaction

$$2 \text{ Bi}_2\text{S}_3 \text{ (s)} + 9 \text{ O}_2 \text{ (g)} \rightarrow 2 \text{ Bi}_2\text{O}_3 \text{ (s)} + 6 \text{ SO}_2 \text{ (g)}$$

The reaction is carried out and kept at 1.00 atm pressure and 175°C throughout. How many liters of O_2 are needed to produce 3.87 L of SO_2? How many grams of Bi_2S_3 are needed? **(E7)**

IV. Gas Mixtures

A. Dalton's Law of Partial Pressures

1. Statement: _____

_____

2. Mathematical relationship

$$P_{total} = P_A + P_B + \ldots$$

3. *Exercises:*

a. A 200.0-mL sample of O_2 is collected over water at 27°C and 748 mm Hg. The vapor pressure of water at 27°C is 28 mm Hg. What mass of O_2 is in the sample?

Solution: To arrive at the mass of O_2, we must first determine the pressure exerted by the oxygen gas alone. To do this we use Dalton's Law of Partial Pressures.

$$P_{total} = P_A + P_B$$
$$= P_{oxygen} + P_{water}$$
$$748 = P_{oxygen} + 28$$
$$P_{oxygen} = 720 \text{ mm Hg}$$

Now, we can calculate the mass of O_2 by using the ideal gas law. The volume of the sample is used as the volume of oxygen.

$$n = \frac{720 \text{ mm Hg} \times \dfrac{1 \text{ atm}}{760 \text{ mm}} \times 0.200 \text{ L}}{0.0821 \dfrac{\text{L} \cdot \text{atm}}{\text{mol} \cdot \text{K}} \times 300 \text{ K}} = 0.00769 \text{ mol}$$

Since 1 mole of O_2 = 32.0 g,

$$0.00769 \text{ mol} \times \frac{32.0 \text{ g}}{1 \text{ mol}} = 0.246 \text{ g } O_2$$

b. Consider the reaction

$$2\,\text{Al (s)} + 6\,H^+ \text{ (aq)} \rightarrow 2\,Al^{3+} \text{ (aq)} + 3\,H_2 \text{ (g)}$$

The hydrogen gas is collected over water at 30.0°C and the total pressure is 740.0 mm Hg (VP_{H_2O} = 32 mm Hg).
(1) What is the partial pressure of the hydrogen gas?

(2) How many grams of H_2 were produced if 2.00 L of wet gas were collected?

(3) How much aluminum metal was used to produce the hydrogen gas? **(E8)**

B. Mole fraction and partial pressure
  1. Mole fraction
     For gas A, it is the fraction of the total number of moles that is accounted for by gas A.

$$X_A = \text{mole fraction of gas A} = \frac{n_A}{n_{total}}$$

n_A = number of moles of gas A

n_{total} = number of moles of all gases combined

  2. Mathematical relationship between the mole fraction and partial pressure

$$P_A = X_A P_{total}$$

  3. *Exercises*
     a. What pressure does a mixture made up of 4.00 g of NO_2 gas and 2.00 g of oxygen gas exert on a 2.00-L container at 27.00°C? What is the partial pressure of the oxygen?
     Solution: To obtain the total pressure, we use the ideal gas law, taking the total number of moles as n. Thus we can write

$$P_{total} \, V = n_{total} \, RT$$

We calculate n_{total} by calculating n_{NO_2} and n_{O_2}

$$n_{O_2} = 2.00 \text{ g} \times \frac{1 \text{ mol}}{32.0 \text{ g}} = 0.0625 \text{ mol O}_2$$

$$n_{NO_2} = 4.00 \text{ g} \times \frac{1 \text{ mol}}{46.0 \text{ g}} = 0.0870 \text{ mol NO}_2$$

Adding we get

$$n_{total} = n_{O_2} + n_{NO_2} = 0.0625 \text{ mol O}_2 + 0.0870 \text{ mol NO}_2 = 0.1495 \text{ mol}$$

Now we substitute n_{total} into the ideal gas law to obtain P_{total}.

$$P_{total} = \frac{0.1495 \text{ mol} \times 0.0821 \frac{\text{L} \cdot \text{atm}}{\text{mol} \cdot \text{K}} \times 300.15 \text{ K}}{2.00 \text{ L}} = 1.84 \text{ atm}$$

Using the relationship

$$P_A = X_A P_{total}$$

where

$$X_A = \frac{n_A}{n_{total}}$$

and $A = O_2$, we get

$$X_{O_2} = \frac{0.0625}{0.0625 + 0.0870} = 0.418$$

and

$$P_{O_2} = 0.418 \times 1.84 = 0.769 \text{ atm}$$

b. Nitrogen gas with a volume of 200.0 mL, a pressure of 750.0 mm Hg, and a temperature of 27.00°C is mixed with oxygen gas and transferred to a 750.0-mL container at 27.00°C. The total pressure of the gas is found to be 680.0 mm Hg, at a temperature of 27.00°C.

(1) Calculate $n_{nitrogen}$.

(2) Calculate n_{total}.

(3) Calculate the partial pressure of each gas. **(E9)**

V. Kinetic Theory of Gases

A. The kinetic theory of gases is based on the idea that _____
_____ .

B. Assumptions of the molecular model
   1. _____
   _____
   2. _____
   _____
   3. _____
   _____
   4. _____
   _____

C. Expression for Pressure
   Mathematical relationship

$$P = \frac{Nmu^2}{3V}$$

   where the ratio N/V expresses _____

   and the product mu^2 is a measure of _____ .

D. Average Kinetic Energy of Translational Motion
   1. Symbol: E_t
   2. Mathematical expression for E_t

$$E_t = \frac{3RT}{2N_A}$$

   3. This means that
      a. At a given temperature, molecules of different gases must have _____
      _____ .

b. E_t is directly proportional to _____

E. Average speed of gas particles

    1. Symbol used for average speed: _____

    2. Equation relating average speed to temperature for any gas.

$$u = \left(\frac{3RT}{MM}\right)^{\frac{1}{2}}$$

    3. The average speed of a gas particle is

       a. directly proportional to _____

       _____

       Equation:

       b. inversely proportional to _____

       _____

       Equation:

    4. Exercises

       a. Calculate the average speed of a neon atom at 27°C.
       Solution: This is a simple plug-in problem. Just remember to use 8.31 × 10^3 g · m^2/s^2 · mol · K for R instead of 0.0821 L·atm/mol·K. Make sure that your units cancel!
       To solve this particular problem, we take the formula

$$u = \left(\frac{3RT}{MM}\right)^{\frac{1}{2}}$$

       and substitute 300 K for T and 20.18 g/mol for MM (molar mass for neon). We get

$$u = \left(\frac{3 \times 8.31 \times 10^3 \, \frac{g \cdot m^2}{s^2 \cdot mol \cdot K} \times 300 \, K}{20.18 \, \frac{g}{mol}}\right)^{\frac{1}{2}} = 609 \, \frac{m}{s}$$

b. At what temperature will an HCl molecule have an average speed of 555 m/s? **(E10)**

c. The average speed of a helium atom at 27°C is 1.37×10^3 m/s.

(*1*) What is the average speed of a gaseous chlorine molecule at 27°C?
Solution: We use the relation

$$\frac{u_{He}}{u_{Cl_2}} = \left(\frac{MM_{Cl_2}}{MM_{He}}\right)^{\frac{1}{2}}$$

Substituting, we get

$$\frac{1.37 \times 10^3 \frac{m}{s}}{u_{Cl_2}} = \left(\frac{70.90}{4.003}\right)^{\frac{1}{2}}$$

$$u_{Cl_2} = \frac{1.37 \times 10^3 \frac{m}{s}}{4.21} = 325 \frac{m}{s}$$

(*2*) What is the average speed of a gaseous helium molecule at 270.0°C?
Solution: This time, we use the relation between speed and temperature for the same molecule. We designate u_2 as the speed at 270.0°C and u_1 as the speed at 27°C.

$$\frac{u_2}{1.37 \times 10^3} = \left(\frac{543}{300}\right)^{\frac{1}{2}}$$

$$u_2 = 1.84 \times 10^3 \frac{m}{s}$$

    d. The average speed of a gaseous HBr molecule at 100°C is 3.39×10^4 m/s.

        (1) What is its speed at 25°C?

        (2) At what temperature could the HBr molecule have an average speed of 7.50×10^5 m/s? **(E11)**

F. Effusion

    1. Definition: _____

_____

    2. Graham's Law

      a. Statement: _____

_____

b. Equations relating the rate of effusion to different variables
(1) To speed of molecules:

(2) To molar mass of the molecules:

(3) To time:

3. *Exercises*

a. An unknown gas goes through a small opening in 65.00 minutes while an equal mass of hydrogen gas goes through the same opening in 9.75 minutes. Calculate the molar mass of the unknown gas.

Solution: Since the variables in this problem are time and mass, we choose the equation that relates them, namely,

$$\frac{\text{time}_A}{\text{time}_B} = \left(\frac{MM_A}{MM_B}\right)^{\frac{1}{2}}$$

If we call the unknown gas A and designate hydrogen as B, then we can write

$$\frac{65.00}{9.75} = \left(\frac{MM_A}{2.02}\right)^{\frac{1}{2}}$$

Squaring both sides we get

$$MM_A = \left(\frac{65.00}{9.75}\right)^2 \times 2.02 = 89.8 \ \frac{g}{mol}$$

   b. What is the molar mass of a gas that diffuses one fifth as fast as helium?
      (E12)

VI. Real Gases
   A. Conditions for deviating from the ideal gas law

   1. _____

   _____

   2. _____

   _____

   B. Factors that the ideal gas law neglects

   1. _____

   _____

   2. _____

   _____

SELF-TEST
A. Multiple choice:
   1. A sealed, rigid flask contains nitrogen gas. The flask is cooled from room temperature
      to $-50°C$. Which of the following statements is true?
      a. The number of moles of nitrogen decreases.
      b. The volume of nitrogen increases.
      c. The pressure of nitrogen decreases.
      d. The pressure of nitrogen increases.
      e. The volume of nitrogen decreases.

   2. The ideal gas law predicts that

      — the volume of a gas goes to zero at absolute zero temperature.
      — density increases with pressure.
      — density increases with temperature.
      — the product, PV/T, for a fixed amount of gas is constant.

   The number of true statements above is
   a. 0   **b.** 1   **c.** 2   **d.** 3   **e.** 4

3. The gas constant, R, can have the units
- bar-m^3/mol-K
- bar-L/mol-K
- mm Hg-cm^3/mol-K
- mol-K/mm Hg-mL

The number of false statements above is
a. 0   **b.** 1   **c.** 2   **d.** 3   **e.** 4

4. In an ideal gas, the collisions of the molecules with the walls of the container account for
a. the velocity of the molecules.
b. the observed pressure.
c. the number of moles.
d. the observed temperature.
e. none of these.

5. For a fixed amount of gas at a fixed pressure, changing the temperature from 30°C to 60°C causes
a. the gas volume to decrease.
b. the gas volume to double.
c. the gas volume to increase but not to double.
d. the gas volume to decrease to half its original volume.
e. no change in the gas volume.

6. Attractive forces between gas molecules are most important at
a. low pressures and low temperatures.
b. low pressures and high temperatures.
c. high pressures and high temperatures.
d. high pressures and low temperatures.

B. Answer the questions below, using **LT** (for *is less than*), **GT** (for *is greater than*), **EQ** (for *is equal to*), or **MI** (for *more information required*) in the blanks provided.

_____ **1.** At 100°C and 1 atm, the velocity of a molecule of hydrogen gas is   (1)   the velocity of a molecule of oxygen gas.

_____ **2.** At 100°C, the average translational kinetic energy of a molecule of hydrogen gas is   (2)   the average translational kinetic energy of a molecule of oxygen gas.

_____ **3.** At constant temperature and volume, the pressure exerted by ten moles of hydrogen gas is   (3)   the pressure exerted by ten moles of oxygen gas.

_____ **4.** At the same temperature and volume, the pressure exerted by 10.00 g of hydrogen gas is ___(4)___ the pressure exerted by 10.00 g of oxygen gas.

_____ **5.** At 50°C, the velocity of a molecule of chlorine gas is ___(5)___ the velocity of the same molecule of chlorine gas at 100°C.

_____ **6.** At a constant volume, the pressure of 2.00 g of SO_2 (g) (M = 64.0 g/mol) at 400 K is ___(6)___ the pressure of 2.00 g of O_2 (g) (M = 32.0 g/mol) at 200 K.

C. True or False:

Consider 2 flasks with the same volume, temperature and pressure. One flask contains nitrogen gas, the other flask has helium gas.

_____ **1.** Both flasks have the same number of moles.

_____ **2.** The nitrogen gas in one flask has a lower density than the helium gas in the other flask.

_____ **3.** If a pinhole is created in both flasks, nitrogen gas would effuse faster than the hydrogen gas.

_____ **4.** The average translational energy of the gases in both flasks is the same.

_____ **5.** One mole of CO_2 gas is introduced to each flask. In the resulting mixture of gases, the mole fraction of nitrogen is equal to the mole fraction of helium.

D. Problems:

Consider ammonia gas (NH_3), which is produced by reacting nitrogen gas and hydrogen gas.

1. Write a balanced equation for the reaction.

2. A one-liter flask contains 0.20 mol of nitrogen at room temperature and 1.5 atm pressure. Another 1.0-L flask contains 0.60 mol of hydrogen at the same room temperature. What is the pressure of the hydrogen gas?

3. The same gases in Problem 2 are combined and put into a 3.0-L flask. The gas mixture exerts a total pressure of 6.0 atm. What temperature does the gas mixture have?

4. The ammonia produced by reacting 0.200 mol of nitrogen and 0.600 mol of hydrogen is measured to be 5.00 L at 27°C and one atm. How many moles of nitrogen were used up in the reaction?

5. Using the data in Problem 4, calculate the partial pressure of each gas in the reaction vessel if all the gases remain in the vessel. The temperature of the vessel is 27°C and the pressure in the vessel is 3.0 atm.

6. Calculate the density of ammonia at 27°C and 1.0 atm pressure.

7. How much faster does the hydrogen diffuse than the nitrogen?

8. Calculate the average speed of an ammonia molecule at 77°C in miles/hour (mph), if oxygen has an average speed of 482 m/s at 25°C.

ANSWERS
Exercises:
(E1) 3.62 bar   **(E2)** 1.090 bar
(E3) 6.22 K   **(E4)** 36.6 g/mol
(E5) 16.0 g/mol
(E6) $2\,Na\,(s) + 2\,H_2O\,(\ell) \rightarrow 2\,Na^+\,(aq) + 2\,OH^-\,(aq) + H_2\,(g)$; 1.14 L
(E7) 5.80 L; 18.0 g   **(E8)** 708 mm Hg; 0.151 g; 1.34 g
(E9) 0.00801 mol;   0.0272 mol; $P_{nitrogen} = 200$ mm; $P_{oxygen} = 480$ mm
(E10) 4.50×10^2 K   **(E11)** 3.03×10^4 m/s; 1.83×10^5 K
(E12) 1.00×10^2 g/mol

Self-Test
A. Multiple Choice:
1. c   2. d   3. b   4. b   5. c   6. b

B. Less than, Equal to, Greater than
1. GT   2. EQ   3. EQ   4. GT   5. LT   6. EQ

C. True or False
1. T   2. F   3. F   4. T   5. T

D. Problems:
1. $N_2 (g) + 3H_2 (g) \rightarrow 2NH_3 (g)$
2. 4.5 atm   3. 2.7×10^2 K   4. 0.101 mol
5. $N_2 = 0.50$ atm; $H_2 = 1.5$ atm; $NH_3 = 1.0$ atm
6. 0.69 g/L   7. 3.72   8. 1.60×10^3 mph

6

Electronic Structure and the Periodic Table

I. Light
  A. Wave Nature of Light
    1. Wavelength

      a. Symbol: _____

      b. Definition: _____

      _____

      c. Units: _____
    2. Frequency

      a. Symbol: _____

      b. Definition: _____

      _____

      c. Units: _____
    3. Mathematical relationship between frequency and wavelength

    Exercises
    a. Microwaves have a wavelength of 1.000 cm. Calculate their frequency in s^{-1}.

    Solution: This problem asks you to relate frequency and wavelength. Hence the equation to use is

$$\nu = \frac{c}{\lambda}$$

Since c is 2.998 × 10⁸ m/s, ν, the wavelength, should be in the same unit of length, m. Thus,

$$\nu = \frac{2.998 \times 10^8 \ \frac{m}{s}}{1.000 \times 10^{-2} \ m} = 2.998 \times 10^{10} \ s^{-1}$$

   b. Calculate the wavelength in nm of a band on your AM radio with a frequency of 4.56×10^5 Hz. **(E1)**

  4. Amplitude

    a. Symbol: _____

    b. Definition: _____

                     _____

B. Particle nature of light
    Two scientists who started to consider light as a stream of particles (photons)

_____ and _____

C. Energy (E) and wave characteristics
  1. Mathematical relationship between energy and
    a. wavelength

    b. frequency

  2. Constants in the mathematical equations
    a. h
      (1) What does it stand for? _____

      (2) What is its value? _____
    b. c
      (1) What does it stand for? _____

      (2) What is its value? _____

3. *Exercises*

a. Calculate the energy of microwaves in joules, using the information obtained in the preceding worked exercise.

Solution: The microwaves in the preceding problem had a frequency of 2.998×10^{10} s^{-1}. We therefore use the relation between frequency and energy

$$E = h\nu$$

where h is Planck's constant. Substituting, we get

$$E = (6.626 \times 10^{-34} \text{ J} \cdot \text{s})(2.998 \times 10^{10} \text{ s}^{-1}) = 1.986 \times 10^{-23} \text{ J}$$

b. Calculate the energy in kJ/mol of an AM radio wave using the information in Exercise E1. **(E2)**

II. The Hydrogen Atom

A. Atomic spectrum of hydrogen

    1. Device that breaks up light into its component colors: _____

    2. Series of lines in the ultraviolet range: _____

    3. Series of lines in the visible range: _____

    4. Series of lines in the infrared range: _____

B. Bohr model of the hydrogen atom

    1. Definition of ground state: _____

    _____

    2. Definition of excited state: _____

    _____

    3. Equation for the energy of the hydrogen atom:

    R_H is _____

    n is _____

    4. Equation relating energy to the level that the electron is in:

5. *Exercises*
  a. Find the energy that is given off in the transition from n = 5 to n = 4.
     Solution: The energy is given by the equation

$$E = R_H \left(\frac{1}{(n_{lo})^2} - \frac{1}{(n_{hi})^2} \right)$$

  We substitute into this equation 5 for n_{hi} and 4 for n_{lo}.

$$E = 2.180 \times 10^{-18} \text{ J} \left(\frac{1}{(4)^2} - \frac{1}{(5)^2} \right) = 4.905 \times 10^{-20} \text{ J}$$

  b. Calculate the wavelength of the light given off in an electron transition be-
     tween two states where the energy difference is 182 kJ/mol. **(E3)**

  c. Calculate the frequency of the light that results from the transition from n =
     6 to n = 3. **(E4)**

C. Quantum mechanical model
   1. Quantum mechanics was developed to describe the motion of small particles
        confined to tiny regions of space.
   2. The quantum mechanical model does not specify the position of the electron

       or the path that it takes about the nucleus. It only deals with _____

       _____ .

   3. _____ developed an equation to calculate the height of the
       electron wave at various points in space.

4. The wave function is given the Greek letter _____ .

5. The square of the wave function is directly proportional to _____

_____ .

6. In the hydrogen atom the electron is more likely to be found _____ .

III. Quantum Numbers and Energy Levels

  A. Definitions

    1. Quantum number: _____

_____

    2. Atomic orbital: _____

_____

  B. First quantum number

    1. Symbol: _____

    2. Values that can be taken on by the first quantum number: _____

    3. The value of the first quantum number designates _____

_____ .

  C. Second quantum number

    1. Symbol: _____

    2. The value of the second quantum number designates _____

_____ .

    3. The second quantum number determines _____

_____ .

    4. Possible values for ℓ: _____
These are always linked to n. The smallest possible ℓ is always 0, and the
largest ℓ is n − 1. Therefore, if you want all possible values for ℓ in the energy
level 1 (n = 1), then you are allowed to go from 0 to (n − 1). In this case, you
go from 0 to (1 − 1), which is also 0. Thus, there is only one possible sublevel
in n = 1 and that is ℓ = 0. What are the possible ℓ values in the principal energy
level n = 4? **(E5)**

5. Designating sublevels in letters instead of numbers

$\ell = 0$ can be designated as the s sublevel

$\ell = 1$ can be designated as the p sublevel

$\ell = 2$ can be designated as the d sublevel

$\ell = 3$ can be designated as the f sublevel

These are all the sublevels you have to know for all the elements that occur naturally plus all those that have been created so far. Memorize these letter designations because you will need to be able to interchange the two designations quickly.

6. Combining n and ℓ

A sublevel is usually referred to in conjunction with the principal energy level that it is in. Thus, a sublevel can be designated as 3p, which means that $n = 3$ and $\ell = 1$. If the designation is 4f, then $n = 4$ and $\ell = 3$.

Warning: Not all combinations will work! Remember that ℓ can only go up to $n - 1$. Thus, 3f is not possible since in 3f, $n = 3$ and $\ell = 3$. For $n = 3$, the maximum ℓ possible is 2 or d. For similar reasons, 1d is not allowed.

7. For multi-electron atoms, energy is dependent on ℓ as on n. Within a given principal level (same value of n) sublevels increase in energy in the order

_____ .

D. Third quantum number

1. Symbol: _____

2. Each sublevel contains one or more _____ which differ from one another in the value assigned to m_ℓ.

3. m_ℓ tells us how the electron cloud surrounding the nucleus is directed in space.

   a. If $\ell = 0$, then $m_\ell = 0$. This is known as the s orbital. The general appearance of the orbital cloud associated with the s orbital is

_____ .

   b. If $\ell = 1$, then m_ℓ can be 1, 0, −1. They are more commonly referred to as p_x, p_y, and p_z orbitals.

   Describe a p orbital: _____

_____

   Orientation of the p orbitals to each other: _____

_____

4. Possible values for m_ℓ:

$$\ell, ...,0...,-\ell$$

5. *Exercises*
   a. If $\ell = 2$, what are the possible m_ℓ values?
      Solution: Since $\ell = 2$ and m_ℓ can go from ℓ to 0 to $-\ell$, the possible values are $-2, -1, 0, 1, 2$.
   b. What are the possible m_ℓ values for the f sublevel? **(E6)**

6. Total number of orbitals per sublevel $= 2\ell + 1$.

E. Fourth quantum number

   1. Symbol: _____
   2. m_s tells us the spin of the electron about its axis.
   3. m_s is not related to n, ℓ, or m_ℓ. It has only two possible values: $+\frac{1}{2}$ or $-\frac{1}{2}$.

F. Pauli exclusion principle
   1. Statement of the principle: _____

   _____

   2. The Pauli exclusion principle and the rules of quantum mechanics fix the capacities of principal levels, sublevels and orbitals. Study the summary given in Table 6.3 of your text.

   3. *Exercises*
      a. Give the quantum numbers for the electrons in the $m_\ell = -2$ orbital of the 3d level.
         Solution: 3d means n = 3 and $\ell = 2$. Thus the quantum numbers are

$$n = 3 \qquad \ell = 2 \qquad m_\ell = -2 \qquad m_s = +\tfrac{1}{2}$$

$$n = 3 \qquad \ell = 2 \qquad m_\ell = -2 \qquad m_s = -\tfrac{1}{2}$$

As you can see, only the spins distinguish these two electrons.
      b. How many electrons are possible in the 4f sublevel?
         Solution: In the 4f sublevel $\ell = 3$; thus the number of possible orbitals is $2\ell + 1$ or 7. Since there are 2 electrons in each orbital, the number of possible electrons in the 4f sublevel is 2(7) = 14.
      c. How many electrons are possible in n = 3? In n = 2, how many electrons are possible with a set of quantum numbers that includes n = 2 and $\ell = 1$? **(E7)**

IV. Electron configurations in the atom
A. From sublevel energies
1. How to write them

a. Determine the number of electrons in the atom by looking up the element's atomic number (Z).

b. Start with $n = 1$ and $\ell = 0$ by writing 1s. Put a superscript above s to designate how many electrons you wish to put there. Remember that for s the maximum superscript is 2. Distribute the electrons, increasing sublevels and energy levels as you have electrons.

c. *Exercises:*

(1) Write the electron configuration for sodium.

Solution: Sodium has atomic number 11. 1s can have 2 for a super-script since you have more than 2 electrons to distribute. Thus, start the electron configuration with $1s^2$, which fills up $n = 1$, since you can only have $\ell = 0 = s$. Then go to $n = 2$, with 2 possible ℓ's, $\ell = 0 = s$ and $\ell = 1 = p$. You therefore write

$$1s^2\, 2s\, 2p$$

2s can again have only a superscript of 2, and since you still have nine electrons to distribute, you can write $1s^2\, 2s^2\, 2p$. Now you have only seven electrons left. The p sublevel can have $m_\ell = -1$, 0, 1, or three orbitals, each orbital having a capacity for two electrons. Thus, 2p can have a maximum superscript of 6. (Using the same reasoning, all p sublevels in any n level can have a maximum superscript of 6.) Since you still have seven electrons, you can write $1s^2\, 2s^2\, 2p^6$. Now, you have only one electron left. So you start with the third energy level, which can have $\ell = 0 = s$, $\ell = 1 = p$, and $\ell = 2 = d$. Capacities for the sublevels and therefore maximum superscripts are $s = 2$, $p = 6$, and $d = 10$. Since you have only one electron left, you only need 3s and the superscript in this case is 1 because that is all you have. Putting it all together, the electron configuration for sodium is

$$1s^2\, 2s^2\, 2p^6\, 3s^1$$

(2) Write the electron configuration for silicon. **(E8)**

2. Sequence of levels and sublevels
   a. The levels and sublevels are written in increasing order of energy. Thus, n = 1 comes before n = 2, and so on.
   b. The sublevels are ordered s < p < d < f.
   c. Ordinarily, sublevels are filled to capacity before the next one starts to fill. This works out quite well for elements with atomic numbers 1 through 18.
   d. After 3p, sublevels start to overlap. Use Figure 6.8 in your text to determine level and sublevel sequences for elements with atomic numbers up to 36. Figure 6.8 gives the order and populations as

$$1s^2 \, 2s^2 \, 2p^6 \, 3s^2 \, 3p^6 \, 4s^2 \, 3d^{10} \, 4p^6$$

3. Abbreviated electron configurations
   a. The abbreviated electron configuration starts with the preceding noble gas, written between brackets.
   b. *Exercises*
      (1) Write the abbreviated electron configuration for cobalt.
         Solution: The preceding noble gas is argon, Ar. We write [Ar] to indicate that the first 18 electrons in the cobalt atom have the electron configuration $1s^2 \, 2s^2 \, 2p^6 \, 3s^2 \, 3p^6$. Since cobalt has 27 electrons, we have nine electrons left. Since 4s comes before 3d and after 3p and since 4s can hold two electrons, our abbreviated electron configuration now looks like this

$$[Ar] \, 4s^2$$

We have seven more electrons to distribute. The next energy level and sublevel available is 3d which can hold a maximum of ten electrons. We therefore put the remaining seven electrons into the 3d sublevel, and the abbreviated electron configuration for cobalt is

$$[Ar] \, 4s^2 3d^7$$

      (2) Write the abbreviated electron configuration for the selenium atom. **(E9)**

B. From the periodic table
   1. The atoms of elements in a group of the periodic table have the same distribution of electrons in the outermost principal energy level, n.
   2. Summary of the order in which electrons fill sublevels using the periodic table.

      a. The elements in Groups 1 and 2 fill _____.

      b. The elements in Groups **13** through **18** fill _____

c. The transition metals fill _____.

d. The lanthanide metals fill _____.

e. The actinide metals fill _____.

3. Figure 6.9 in your text can be used to deduce the electron configuration of any element.

4. *Exercises*

   a. Write the electron configuration for antimony (Sb).

   Solution: We look for antimony in the periodic table and find that it is past period 1, so the 1s sublevel is filled: $1s^2$.

   Antimony is past period 2 in which the 2s and 2p sublevels are filled: $2s^2\,2p^6$.

   Antimony is past period 3 in which the 3s and 3p sublevels are filled: $3s^2\,3p^6$.

   Antimony is past period 4 in which the 4s, 3d, and 4p (in that order) sublevels are filled: $4s^2\,3d^{10}\,4p^6$.

   Antimony is in the fifth period, past Groups 1 and 2, thus the 5s sublevel is filled ($5s^2$), and past the transition metal groups (Groups 3-12) so the 4d sublevel is filled ($4d^{10}$). Antimony is in Group **15**, which means that there are 3 electrons in the 5p sublevel ($5p^3$).

   Putting it all together, the electron configuration for antimony is

   $$1s^2\,2s^2\,2p^6\,3s^2\,3p^6\,4s^2\,3d^{10}\,4p^6\,5s^2\,4d^{10}\,5p^3$$

   Check your work by adding up the superscripts. Your total should equal the atomic number for antimony, which it does.

   b. Write the electron configuration for lead. **(E10)**

V. Orbital Diagrams of Atoms

A. Setting up diagrams

1. You will need to know the electron configuration of the atom for which you want to write an orbital diagram.

2. *Exercise*

   Set up the orbital diagram for sulfur.

   The electron configuration for sulfur is $1s^2\,2s^2\,2p^6\,3s^2\,3p^4$. Drawing a pair of parentheses to represent each orbital in the sublevel, you find that the setup will look like this:

$1s^2$	$2s^2$	$2p^6$	$3s^2$	$3p^4$
( )	( )	( )( )( )	( )	( )( )( )

B. Filling in the orbitals
 1. Hund's rule
  There is no problem in filling in the orbitals in the s sublevels. Hund's rule comes into play when filling orbitals in the p, d, and f sublevels, when they are not filled to capacity. In the case of sulfur, filling the orbitals is straightforward until we come to the 3p sublevel. Hund's rule says that all orbitals are first filled with one electron each. Then if there are more electrons to go around, the orbitals can be filled with the second electron. The orbital diagram for sulfur is

$$(\uparrow\downarrow) \quad (\uparrow\downarrow) \quad (\uparrow\downarrow)(\uparrow\downarrow)(\uparrow\downarrow) \quad (\uparrow\downarrow) \quad (\uparrow\downarrow)(\uparrow\ )(\uparrow\ )$$

    1s 2s   2p   3s   3p

 2. Exercises
  a. Write the orbital diagram for cobalt. **(E11)**

  b. Write the orbital diagram for fluorine. **(E12)**

VI. Electron Configurations in Monatomic Ions
 A. Ions with noble gas structures.
  1. Main-group elements form ions that have the same number of electrons as the noble gas atom.
  2. Note that the noble gas atom closest to the element does not necessarily have to be the preceding noble gas. Sodium, for example, loses its $3s^1$ electron to achieve the configuration of Ne, which precedes it, but fluorine gains an electron to achieve the configuration of neon, which follows it in the periodic table.

  3. Species with the same electron configuration are said to be _____.
 B. Transition metal cations
  1. The transition metals (Groups 3-12) do not form ions with noble gas configurations.
  2. Which electrons are lost when transition metal atoms form cations with +1 or +2 charges?

   _____

  3. Which electrons are lost when transition metal atoms form cations with +3 or +4 charges?

   _____

4. *Exercises*

a. Write the electron configuration for As^{3-} and for Cr^{3+}.

Solution: (1) Arsenic is a main-group metal, so it forms the electron configuration of its nearest noble gas, which is Kr. Its electron configuration is

$$1s^2\, 2s^2\, 2p^6\, 3s^2\, 3p^6\, 4s^2\, 3d^{10}\, 4p^6$$

(2) Chromium is a transition metal. Its cation has a charge of +3 so it loses both of its outermost s electrons first ($4s^2$). It also loses one of its outermost d electrons (3d). The electron configuration for the Cr^{3+} is

$$1s^2\, 2s^2\, 2p^6\, 3s^2\, 3p^6\, 3d^3$$

To check, Cr metal has 24 electrons. Cr^{3+} lost three electrons, so there must be 21 electrons left. Adding the superscripts in the electron configuration for Cr^{3+} gives 21 as a total.

b. Write the abbreviated electron configuration for Fe^{2+}, Fe^{3+}, and Te^{2-}. **(E13)**

VII. Periodic Trends in the Properties of Atoms

A. Atomic radius

1. Atomic radius is a measure of _____.

2. Atomic radius decreases as we move across the periodic table from left to right and increases as we move down within a group.

3. Effective nuclear charge

Effective nuclear charge is the number of electrons in the outer electron configuration. Sodium's outer electron configuration is $3s^1$. Thus, its effective nuclear charge is 1.

4. As we go down a group, the effective nuclear charge does not change, but the energy levels (or layers) increase, so size increases. Going across the periodic table, the number of layers is constant, but the effective nuclear charge increases. Thus electrons are pulled in toward the nucleus, and size decreases.

B. Ionic radius

1. Just as with their corresponding atoms, ionic radii increase as we move down within a group. The ionic radii increase as we move across a period.

2. Cations are smaller than their corresponding atoms, because in the cation there are more protons than electrons. Therefore, the electrons are attracted and move closer to the nucleus.

3. Anions are larger than their corresponding atoms, because the presence of the extra electrons causes the electrons to repel each other. As a result, they try to move as far away from each other as possible, creating bigger clouds.

C. Ionization energy
1. It is a measure of _____

   _____ .

2. The smaller the ionization energy is, the easier the electron is to remove.
3. As we go across the periodic table from left to right, the distance between the nucleus and the electrons decreases, the electrons become more tightly bound, and the ionization energy increases. As we move down a group, the electrons are separated from the nucleus by more layers. Consequently, the nucleus exerts less attraction on the electrons, and it becomes easier to pull out an electron. So ionization energy gets smaller.

D. Electronegativity
1. It is a measure of _____

   _____ .

2. The greater the electronegativity, the greater affinity the atom has for electrons.
3. Electronegativity _____ as one moves from left to right in the periodic table.
4. Electronegativity _____ as one moves down a group in the periodic table.

SELF-TEST
A. Multiple Choice:
1. How many unpaired electrons are there in an atom with the electron configuration

   $$1s^2\, 2s^2\, 2p^6\, 3s^2\, 3p^6\, 4s^2\, 3d^{10}\, 4p^6$$

   a. 1   **b.** 2   **c.** 3   **d.** 4   **e.** some other number

2. Which one of the following represents a possible *excited* electron configuration, in which an electron has been promoted from the ground state to a higher level?
   a. $1s^2\, 2s^2\, 2p^6\, 3s^2$ **b.** $1s^2\, 2s^2\, 2p^6\, 3s^1\, 3p^1$ **c.** $1s^2\, 2s^2\, 2p^6\, 3s^3$
   d. $1s^2\, 2s^2\, 2p^5\, 2d^1\, 3p^2$ **e.** $1s^2\, 2s^2\, 2p^7\, 3s^1$

3. Which of the following statements is correct?
   a. Atomic radii increase as one moves from left to right across a period.
   b. Atomic radii increase as one moves down a group.
   c. Negative ions are smaller than the nonmetals from which they are formed.
   d. Positive ions are larger than the metals from which they are formed.

4. In which of the following series of atoms are the first ionization energies decreasing?
   a. Sr > Ba > Ra   **b.** Li > Be > B
   c. O > F > S   **d.** As > P > N

5. The maximum number of electrons possible in an f sublevel is
   a. 2   **b.** 6   **c.** 14   **d.** 32

6. There are two very intense lines in the atomic spectrum of sodium. The wavelengths are 589.0 nm and 589.6 nm. Which line corresponds to the larger energy?
   a. Energy is the same for both lines.
   b. Energy is greater for the 589.0 nm line.
   c. Energy is greater for the 589.6 nm line.
   d. Cannot answer because insufficient information is given.

7. How many elements in the second period of the periodic table (starting with Li) have one or more unpaired electrons?
   a. 3   **b.** 4   **c.** 5   **d.** 6   **e.** 7

8. A filled d sublevel and a half-filled f sublevel contain a *total* of how many electrons?
   a. 11   **b.** 17   **c.** 19   **d.** 24

9. Given the following orbital diagram

	1s	2s	2p	
—	(↑)	(↑)	( )( )( )	= hydrogen atom – excited state
—	(↑↓)	( )	( )( )( )	= hydrogen atom – ground state
—	(↑↓)	( )	( )( )( )	= helium atom – ground state
—	(↑)	(↑)	( )( )( )	= helium atom – excited state

   The number of correct orbital diagrams is
   a. 0   **b.** 1   **c.** 2   **d.** 3   **e.** 4

10. The radius of

    (1) Br^- > Br   **(2)** Li < Na   **(3)** Mg > Cs   **(4)** K^+ > Cs^+

    The true statements above are
    a. (1),(2)   **b.** (1),(3)   **c.** (2),(3)   **d.** (2),(4)   **e.** (3),(4)

B. True or False

_____ 1. The energy required to make the transition from n = 1 to n = 3 is twice the energy required to make the transition from n = 1 to n = 2.

_____ 2. A 3p subshell can have an electron with quantum number $\ell = 2$.

_____ 3. An important factor in determining the relative size of halogen atoms in their ground state is the principal quantum number.

_____ 4. The first two quantum numbers (n, ℓ) of the highest energy electron in the ground state of Sc^{3+} are 3,1.

_____ 5. The electron configuration $1s^2\ 2s^2\ 2p^4\ 2d^1$ may be that of fluorine in an excited state.

_____ 6. Pauli's exclusion principle states that two electrons in the same orbital have the same spin.

_____ 7. It is possible for an electron to have $4,0,1,\frac{1}{2}$ as its set of 4 quantum numbers.

_____ 8. The ionic radius of Na^+ is larger than the atomic radius of Na.

C. Fill in the Blanks

_____ 1. How many electrons with quantum number $\ell = 1$ are there in the ground state electron configuration for potassium?

_____ 2. How many electrons can have both quantum numbers n = 2 and $\ell = 1$?

_____ 3. How many unpaired electrons are there for arsenic in its ground state?

_____ 4. What is the abbreviated ground state electron configuration for Zr^{3+}?

_____ 5. What neutral element is represented by the electron configuration $1s^2\ 2s^2\ 2p^6\ 3s^1\ 3p^1$?

_____ 6. Does the electron configuration in #5 represent the element in its ground state?

_____ 7. What is the symbol of the most electronegative element in Group 15?

_____ 8. Which neutral element in Period 4 has the largest radius?

_____ **9.** Which neutral element in Group 2 has the largest first ionization energy?

_____ **10.** Write the abbreviated ground state electron configuration for nickel.

D. Answer the questions below, using **LT** (for *less than*), **GT** (for *greater than*), **EQ** (for *equal to*), or **MI** (for *more information required*) in the blanks provided.

_____ **1.** The wavelength of the photon required to promote an electron in the hydrogen atom from the $n = 1$ to the $n = 3$ level is ___(1)___ the wavelength of the photon required to promote an electron in the hydrogen atom from the $n = 1$ to the $n = 2$ level.

_____ **2.** For a scandium atom in the ground state, the energy of the 3d sublevel is ___(2)___ the energy of the 4s sublevel.

_____ **3.** The number of unpaired electrons for a carbon atom in the ground state is ___(3)___ the number of unpaired electrons for an oxygen atom in the ground state.

_____ **4.** The ℓ quantum number of the 3p sublevel is ___(4)___ the ℓ quantum number of the 5p sublevel.

_____ **5.** The atomic radius of the fluorine atom is ___(5)___ the atomic radius of the fluoride ion.

_____ **6.** The atomic radius of the magnesium atom is ___(6)___ the atomic radius of the chlorine atom.

_____ **7.** The electronegativity of the halogen atom in the second period of the periodic table is ___(7)___ the electronegativity of the halogen atom in the fourth period of the periodic table.

_____ **8.** The first ionization energy of potassium is ___(8)___ the first ionization energy of calcium.

E. Problems:

Consider the second line of the Paschen series (n = 5 to n = 3) of the hydrogen atom.

1. Calculate ΔE in kJ/mol.

2. Calculate λ.

3. Calculate ν.

Consider technecium (Z = 43).

4. Write its electron configuration.

5. Write its orbital diagram.

6. Write its abbreviated electron configuration.

7. Write the abbreviated electron configuration for Tc^{2+}.

A new element was made and found to have 115 protons and 187 neutrons,

8. Its atomic number is _____.

9. Its outer electron configuration is _____.

10. The element it most resembles in chemical properties is _____.

11. Compare its atomic radius with that of Fr.

12. Compare its ionization energy with that of As.

13. Compare its electronegativity with that of Po.

ANSWERS
Exercises:
(E1) 6.57×10^{11} nm   **(E2)** 1.82×10^{-7} kJ/mol   **(E3)** 657 nm
(E4) 2.742×10^{14} s^{-1} **(E5)** 0, 1, 2, 3   **(E6)** 3, 2, 1, 0, -1, -2, -3
(E7) 18; 6   **(E8)** $1s^2 2s^2 2p^6 3s^2 3p^2$ **(E9)** [Ar] $4s^2 3d^{10} 4p^4$
(E10) $1s^2 2s^2 2p^6 3s^2 3p^6 4s^2 3d^{10} 4p^6 5s^2 4d^{10} 5p^6 6s^2 4f^{14} 5d^{10} 6p^2$
(E11) (↑↓)   (↑↓)   (↑↓)(↑↓)(↑↓)   (↑↓)   (↑↓)(↑↓)(↑↓)   (↑↓)   (↑↓)(↑↓)(↑)(↑)(↑)
(E12) (↑↓)   (↑↓)   (↑↓)(↑↓)(↑)
(E13) [Ar] $3d^6$, [Ar] $3d^5$, [Xe]

Self-Test
A. Multiple Choice:

1. e	**2.** b	**3.** b	**4.** a	**5.** c
6. b	**7.** d	**8.** b	**9.** c	**10.** a

B. True or False:

1. F	**2.** F	**3.** T	**4.** T	**5.** F
6. F	**7.** F	**8.** F		

C. Fill in the Blanks:

1. 12	**2.** 6	**3.** 3	**4.** [Kr] $4d^1$	**5.** Mg
6. No	**7.** N	**8.** K	**9.** Be	**10.** [Ar] $4s^2 3d^8$

D. Less than, Equal to, Greater than:

1. LT	**2.** GT	**3.** EQ	**4.** EQ	**5.** LT
6. GT	**7.** GT	**8.** LT		

E. Problems:

1. 93.35 kJ/mol   2. 1.281×10^3 nm   3. 2.340×10^{14} s^{-1}

4. $1s^2\, 2s^2\, 2p^6\, 3s^2\, 3p^6\, 4s^2\, 3d^{10}\, 4p^6\, 5s^2\, 4d^5$

5. (↑↓)   (↑↓)   (↑↓)(↑↓)(↑↓)   (↑↓)   (↑↓)(↑↓)(↑↓)   (↑↓)   (↑↓)(↑↓)(↑↓)(↑↓)(↑↓)
    1s   2s   2p   3s   3p   4s   3d

   (↑↓)(↑↓)(↑↓)   (↑↓)   (↑)(↑)(↑)(↑)(↑)
       4p   5s   4d

6. [Kr] $5s^2\, 4d^5$ 7. [Kr] $4d^5$ 8. 115

9. [Rn] $7s^2\, 7p^3$ 10. Bi   11. smaller

12. smaller   13. smaller

7

Covalent Bonding

I. Lewis Structures
A. Nomenclature
1. Valence electrons

   a. Definition: _____

      _____

   b. It is easy to determine the number of valence electrons for a main-group element. The number of valence electrons is the same as the second digit of the group number for non-transition elements. Since Groups 1 and 2 have only one digit in their group number, that is also the number of valence electrons for the elements in those groups.

   c. Valence electrons are represented by dots in a Lewis structure, each dot representing one electron.

   d. Exercises

      (1) How many valence electrons are there in the oxygen atom?

         Solution: Since oxygen is in Group **16**, there are six valence electrons.

      (2) How many valence electrons are there in the aluminum atom? **(E1)**

   e. Notes

     (1) Do not confuse valence electrons with valence. You might have heard about valences in high school. Valences have very little to do with valence electrons, and it is advisable to remove the noun "valence" from your vocabulary. From now on, when you hear or see the word "valence", it should be the adjective for the noun "electron".

     (2) The number of valence electrons is always positive and never more than eight. There is no such thing as a negative number of electrons.

2. Unshared electrons or lone pairs

Definition: _____

_____

3. Covalent bonds

   a. When two electrons are shared by two nuclei, a covalent bond is formed. This bond is represented in structures by a straight line.

   b. Two nuclei may share more than two electrons. If four electrons are shared, then a double bond is formed. If six electrons are shared, a triple bond is formed.

4. Octet rule

Statement of the rule: _____

_____

B. Writing Lewis structures

   1. Rules

   a. _____

   _____

   b. _____

   _____

   c. _____

   _____

   d. _____

   _____

   2. Notes on writing skeleton structures

   a. When you have three or more different atoms in a molecule, several factors can influence your choice for the central atom.

   (1) If carbon is one of the atoms, it is always the central atom. Hydrogen, oxygen, and the halogens are almost always terminal atoms.

   (2) The atom whose position in the periodic table is farthest away from the right edge is usually the central atom because it is the least electronegative atom.

   b. Exercises

   (1) Write the skeleton structure for CH_2Cl_2.

   Solution: Carbon is in Group 14, hydrogen in Group 1, and chlorine in Group 17. Thus, carbon is the farthest away from the edge of the periodic table and can be designated as the central atom. The skeleton structure for CH_2Cl_2 is then

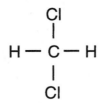

The spatial arrangement of the atoms attached to the central atom is not important for the skeleton structure. Therefore, it is just as good to write

$$
\begin{array}{c}
\text{Cl} \\
| \\
\text{H} - \text{C} - \text{Cl} \\
| \\
\text{H}
\end{array}
$$

(2) Write the skeleton structure of NOCl. **(E2)**

    c. Consider the case of a molecule of the form XY_n, where you have one atom of the element X and several atoms of the element Y. The skeleton structure for this type of molecule ordinarily has the single atom element as the central atom.

       Example: Consider the polyatomic ion ClO_4^-. Even though Cl is closer to the edge of the periodic table than is O, Cl is the central atom because there is only one atom of Cl.

3. Note on counting valence electrons:
If the total number of valence electrons that you get for an answer is odd (that is, 7, 9,...), try again. Odd numbers of total valence electrons are highly unlikely.

4. Note on the number of bonds:
The skeletal structure always starts out with single bonds. If, after distributing the electrons, you are short, then you can move an electron to form another bond. Remember, after you have drawn your skeleton structure, first try to distribute electrons. Add more bonds only at the end.

5. *Exercises*
    a. Write the Lewis structure for ClO_4^-.
      Solution:
      (1) The skeleton structure for ClO_4^- is

$$\begin{array}{c} O \\ | \\ O - Cl - O \\ | \\ O \end{array}$$

(2) Chlorine is in Group **17**, thus it has 7 valence electrons. Oxygen is in Group **16**, thus each oxygen atom has 6 valence electrons. Four oxygen atoms therefore have 24 valence electrons. Since there is a charge of −1, we have to add one more electron to the number of valence electrons. There are 7 + 24 + 1 = 32 valence electrons.

(3) Since there are four single bonds in the skeleton structure, we subtract 8 from the total number of valence electrons, leaving 24 electrons left to distribute. The chlorine atom has its octet with the four bonds (8 electrons) around it. Each oxygen atom has a bond (2 electrons), so 6 electrons per oxygen atom are needed. There are 24 electrons left to distribute, so each oxygen atom will have an octet. The Lewis structure is therefore

$$\left(\begin{array}{c} :\ddot{O}: \\ | \\ :\ddot{O} - Cl - \ddot{O}: \\ | \\ :\ddot{O}: \end{array} \right)^{-}$$

b. Write the Lewis structure for PH_3. **(E3)**

c. Write the Lewis structure for $(NO_2)^{-}$. **(E4)**

C. Resonance forms

1. When writing resonance forms, you can only shift electrons, not atoms. In other words, stick to your skeleton structure. All resonance forms of the same molecule have the same skeleton structure.

2. Give all the resonance forms of SO_3. **(E5)**

D. Formal charge

1. Defining equation

$$C_f = X - \left(Y + \frac{Z}{2}\right)$$

where

$C_f =$ _____

$X =$ _____

$Y =$ _____

$Z =$ _____

2. Criteria to follow in choosing the most likely Lewis structure:

   a. _____

   _____

   b. _____

   _____

3. Exercises

   a. There are two possible Lewis structures for HCN. They are

$$H - C \equiv N :   \quad \text{or} \quad   H - N \equiv C :$$

Using the concept of formal charge, decide which is the more likely structure. *Solution:* We determine the formal charge of each structure by using the formula

$$C_f = X - \left(Y + \frac{Z}{2}\right)$$

For $H - C \equiv N:$

$$C_f \text{ of } C = 4 - \left[0 + \frac{8}{2}\right] = 4 - 4 = 0$$

$$C_f \text{ of } N = 5 - \left[2 + \frac{6}{2}\right] = 5 - 5 = 0$$

For $H - N \equiv C:$

$$C_f \text{ of } C = 4 - \left[2 + \frac{6}{2}\right] = 4 - 5 = -1$$

$$C_f \text{ of } N = 5 - \left[0 + \frac{8}{2}\right] = 5 - 4 = 1$$

Clearly, the more likely structure is $H - C \equiv N:$ because all the formal charges in this structure are equal to zero. Furthermore, the negative formal charge on the second structure is on the carbon atom. It is less electronegative than nitrogen, which has a positive formal charge.

b. Using the concept of formal charge, choose the more likely skeleton structure for N_2O: N–N–O or N–O–N. **(E6)**

E. Exceptions to the octet rule
  1. Electron-deficient species (free radicals)
     a. Examples:

        _____

     b. Paramagnetic – definition: _____

        _____

     c. The fluorides of Be and B
     d. Memorize the Lewis structures of NO, NO_2, BeF_2, and BeF_3.
  2. Expanded octets
     a. Atoms in expanded octets
        (1) The central atom is a non-metal from Groups 15 - 18. It must have d orbitals available. Thus, even though N is a non-metal in Group 15, it cannot be a central atom for an expanded octet because it does not have any d orbitals available.

(2) Terminal atoms are most often the halogens. Occasionally, oxygen atoms are present as terminal atoms.

b. Drawing the Lewis structure

(1) Proceed in the same manner as for regular octets.

(2) Extra electrons are always distributed around the central atom. Terminal atoms still follow the octet rule. Only the central atom violates it.

(3) Exercises

(a) Write the Lewis structure for ClF_3.

Solution: Following the steps for writing the Lewis structure of an octet, we first count the number of valence electrons.

7 for Cl and 3(7) for F which total 28

The skeleton structure is

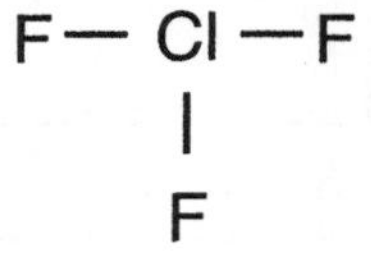

$$F - Cl - F$$
$$|$$
$$F$$

The number of valence electrons for distribution is

$$28 - 3(2) = 22$$

We only need, however, 20 electrons to complete all octets. Extra electrons are always given to the central atom, in this case, Cl. Thus we have a structure where Cl has 10 electrons around it.

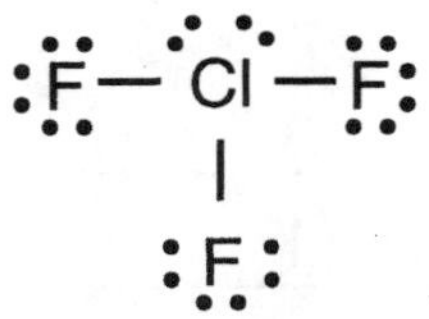

$$:\!\overset{..}{\underset{..}{F}} - \overset{..}{Cl} - \overset{..}{\underset{..}{F}}\!:$$
$$|$$
$$:\!\overset{}{\underset{..}{F}}\!:$$

(b) Write the Lewis structure for XeF_2. **(E7)**

II. Molecular Geometry

A. Definitions

1. Bond angle: _____

_____

2. VSEPR: _____

3. VSEPR theory: _____

_____

B. Ideal geometries
1. Species with ideal geometries are those with a central atom surrounded by from two to six electron pairs, all of which form single bonds with terminal atoms. There are no unshared pairs around the central atom.
2. Two single bonds

Predicted geometry is _____.

Bond angle is _____.

Example: _____
3. Three single bonds

Predicted geometry is _____.

Bond angle is _____.

Example: _____
4. Four single bonds

Predicted geometry is _____.

Bond angle is _____.

Example: _____
5. Five single bonds

Predicted geometry is _____.

Bond angle is _____.

Example: _____
6. Six single bonds

Predicted geometry is _____.

Bond angle is _____.

Example: _____

C. Species with unshared pairs around the central atom
1. Two single bonds and one unshared pair

Predicted geometry is _____.

Bond angle is _____.

Example: _____
2. Two single bonds and two unshared pairs

Predicted geometry is _____.

Bond angle is _____.

Example: _____

3. Two single bonds and three unshared pairs

   Predicted geometry is _____.

   Bond angle is _____.

   Example: _____

4. Three single bonds and one unshared pair

   Predicted geometry is _____.

   Bond angle is _____.

   Example: _____

5. Three single bonds and two unshared pairs

   Predicted geometry is _____.

   Bond angle is _____.

   Example: _____

6. Four single bonds and one unshared pair

   Predicted geometry is _____.

   Bond angle is _____.

   Example: _____

7. Four single bonds and two unshared pairs

   Predicted geometry is _____.

   Bond angle is _____.

   Example: _____

8. Five single bonds and one unshared pair

   Predicted geometry is _____.

   Bond angle is _____.

   Example: _____

D. Multiple bonds

For purposes of predicting geometry, multiple bonds are counted as a single electron pair. Hence, a double bond is counted as a single bond. For example, acetylene (C_2H_2) has the Lewis structure

$$H - \overset{*}{C} \equiv C - H$$

The geometry of the starred carbon atom can be predicted by counting two single bonds. The H–C counts as one of the single bonds and the $C \equiv C$ as another. Thus its geometry is linear and its approximate bond angle is 180°.

E. Notes on geometry

    1. You cannot predict the geometry of a molecule unless you know its Lewis structure.

    2. Remember, the bonds and unshared pairs of electrons that you count are only those attached to the central atom.

    3. It might be easier to remember that the bond angle is dependent only on two things — the number of unshared pairs around the central atom and the number of atoms bonded to the central atom. Thus,

> number of unshared pairs + number of atoms
> = number of electron pairs around the central atom

If there are four electron pairs around the central atom, the molecule will have an approximate bond angle of 109°.

If the molecule has three electron pairs around the central atom, its bond angle is approximately 120°.

If the molecule has two electron pairs around the central atom, its bond angle is approximately 180°.

F. *Exercises*

    1. Predict the geometry and the approximate bond angle for H_3O^+.

    Solution: The Lewis Structure of the molecule is

$$\left(H - \overset{\cdot\cdot}{O} - H \atop \underset{H}{|} \right)^{+}$$

There are four electron pairs (1 unshared pair + 3 atoms bonded) around the central atom O and so the bond angle is 109°. Three of the four electron pairs are used as single bonds and one is unshared. Thus the geometry is pyramidal.

    2. Predict the geometry of the following species. Give the bond angle of the first three species.

    a. HCN **(E8)**

    b. SO_3 **(E9)**

         c. NH_4^+ **(E10)**

         d. $SbCl_5$ **(E11)**

         e. $SnCl_6^{2-}$ **(E12)**

III. Polarity
  A. Polarity in covalent bonds
     1. Definitions
       a. Polar bond: _____

                  _____

       b. Dipole: _____

                  _____

       c. Nonpolar bond: _____

                  _____

     2. A nonpolar bond is formed whenever the two atoms joined together by the bond are identical.

3. Covalent bonds between two unlike atoms are always polar.

4. The extent of the polarity of a covalent bond is related to _____.

B. Polarity of molecules

    1. Nomenclature

        a. Polar molecule — definition:

        _____

        _____

        b. Nonpolar molecule — definition:

        _____

        _____

    2. Polarity dependence on geometry

        a. Molecules with tetrahedral, triangular planar, and linear geometries can be polar or nonpolar. All molecules with bent (either with a 109° bond angle or a 120° bond angle) or pyramidal geometries are polar. The presence of an unshared pair of electrons almost always makes the molecule polar.

        b. In determining the polarity of molecules with symmetric geometries (linear, triangular planar, or tetrahedral) it is important to note the identity of the atoms attached to the central atom. In the case of molecules whose geometries are linear and triangular planar, the atoms around the central atom have to be identical for the molecule to be nonpolar.

    3. *Exercises*

        a. Predict the polarity of H_3O^+.

        Solution: The Lewis structure of the molecule is

$$\left(\begin{array}{c} H - \ddot{O} - H \\ | \\ H \end{array} \right)^{+}$$

        The geometry is pyramidal. Hence the molecule has to be polar.

        b. Predict the polarity of the following molecules.

            (1) HCN **(E13)**

            (2) SO_3 **(E14)**

(3) NH_4^+ **(E15)**

IV. Hybridization; Multiple bonds

A. Notes

   1. Hybridization, like bond angle, is dependent on the number of electron pairs around the central atom. If you want to determine a molecule's hybridization, simply count all bonded atoms and unshared pairs around the central atom. Remember, double bonds and triple bonds count as a single bond.

   2. Molecules that obey the octet rule use s and p orbitals as hybrid orbitals. Thus a total of four bonds and unshared pairs has an sp^3 hybridization. A total of three gives hybridization of sp^2, while a total of two gives sp hybridization.

   3. Expanded octets have hybridized d orbitals in addition to the s and p orbitals. Thus a total of five bonds and unshared pairs has an sp^3d hybridization, while a total of six gives an sp^3d^2 hybridization.

B. Sigma (σ) and pi (π) bonds

   1. All single bonds are sigma bonds.

   2. If the molecule has a double bond, one of the bonds is sigma and the other is pi.

   3. If the bond has a triple bond, one of the bonds is sigma and the other two are pi.

   4. Remember that only in counting sigma and pi bonds do you count all three bonds in the triple bond and both bonds in the double bond.

C. *Exercises*

   1. Give the hybridization, number of sigma bonds, and number of pi bonds for H_3O^+.

   Solution: The Lewis structure for the molecule is

$$\left(\begin{array}{c} H - \overset{\cdot\cdot}{O} - H \\ | \\ H \end{array} \right)^{+}$$

   Since there are four electron pairs around the central atom (three bonded atoms and one unshared pair), the hybridization is sp^3. There are three single bonds, so there are three sigma bonds and no pi bonds.

   2. Give the hybridization of the central atom, the number of sigma bonds, and the number of pi bonds for the following molecules.

   a. HCN **(E16)**

b. SO_3 **(E17)**

c. NH_4^+ **(E18)**

d. $SbCl_5$ **(E19)**

e. $SnCl_6^{2-}$ **(E20)**

SELF-TEST
Multiple Choice:
1. The total number of electron pairs around phosphorus for PCl_5 is
   a. 2   **b.** 3   **c.** 4   **d.** 5   **e.** 6

2. An atom X is surrounded by 3 sigma bonds and 1 pi bond. The approximate bond
   angle around the central atom is
   a. 90°   **b.** 109°   **c.** 120°   **d.** 180°

3. An atom X is surrounded by an unshared pair of electrons, 2 sigma bonds, and one
   pi bond. The hybridization for the molecule with X as central atom is
   a. sp   **b.** sp^2 **c.** sp^3 **d.** dsp^3 **e.** d^2sp^3

4. Which of the following bonds would be the most polar?
   a. C—C   **b.** C—O   **c.** C—F   **d.** C—N

5. All of the following molecules have double bonds except
   a. BeF_2 **b.** CO_2 **c.** SO_3 **d.** NO

6. Chlorine pentafluoride forms an expanded octet. Which of the following statements
   are true?

   (1) There are 6 bonds to the central atom.
   (2) There is a pair of unshared electrons around the central atom.
   (3) Its geometry is that of a square pyramid.
   (4) Its geometry is tetrahedral.

   a. (1),(3)   **b.** (2),(4)   **c.** (1),(4)   **d.** (2),(3)

7. Which of the following form expanded octets?

   (1) NF_3 **(2)** CO_3^{2-} **(3)** XeF_4 **(4)** ClF_3
   a. (1),(2)   **b.** (3),(4)   **c.** (1),(4)   **d.** (2),(3)

8. For which of the following species can one write reasonable resonance structures?

   (1) ClO_2^- **(2)** CO_3^{2-} **(3)** SO_2 **(4)** NO_2^- **(5)** ClO_4^-
   a. (1),(2),(3)   **b.** (2),(3),(4)   **c.** (3),(4),(5)   **d.** (2),(3)   **e.** (3),(5)

9. Which of the following lists the molecular geometries of ClF_5, SO_3, and PF_5 correctly
   and in order?
   a. square pyramid, triangular planar, triangular bipyramid
   b. octahedral, bent, triangular bipyramid
   c. octahedral, triangular planar, triangular bipyramid
   d. triangular bipyramid, linear, tetrahedral

10. Which of the following molecules is/are nonpolar?

    a. $(XeF_5)^+$ **b.** CH_2F_2 **c.** SO_2 **d.** OF_2

    e. a, b, c, d are polar   **f.** a, b, c, d are nonpolar

Problems:

Give the Lewis structures and all resonance forms, if any, of the following:

  1. PO_3^-

  2. CS_2

  3. ClO_3^-

Consider the following molecules. Determine
   a. their Lewis structures
   b. their geometry
   c. their bond angles
   d. their polarity
   e. the hybridization of the central atom
   f. the number of sigma and pi bonds

If the molecule is an expanded octet, determine only its Lewis structure, geometry, and hybridization.

   4. OCS

   5. NO_3^-

   6. H_3C—O—CCl_3

   7. $(SbF_5)^{2-}$

8. PCl_3F_2

ANSWERS
Exercises:

(E1) 3

(E2) Cl—N—O

(E3)   H—P̈—H
              |
              H

(E4) $\left(:\ddot{O}-\ddot{N}=\ddot{O}:\right)^{-}$

(E5) :Ö—S—Ö: ⟷ :Ö=S—Ö: ⟷ :Ö—S=Ö:
              ‖ | |
             :O: :O: :O:

(E6) N≡N—O

(E7) :F̈—Xe—F̈:

(E8) linear, 180°

(E9) triangular planar, 120°

(E10) tetrahedral, 109°

(E11) triangular bipyramid

(E12) octahedral

(E13) polar

(E14) nonpolar

(E15) nonpolar

(E16) sp, 2σ, 2π

(E17) sp^2, 3σ, 1π

(E18) sp^3, 4σ

(E19) sp^3d, 5σ

(E20) sp^3d^2, 6σ

Self-Test
Multiple Choice:

1. d   **2.** c   **3.** b   **4.** c   **5.** a   **6.** d   **7.** b   **8.** b   **9.** a   **10.** e

Problems:

1. $\left(\ddot{O}=P-\ddot{O}: \quad :\ddot{O}: \right)^{-}$ ⟷ $\left(:\ddot{O}-P=\ddot{O} \quad :\ddot{O}: \right)^{-}$ ⟷ $\left(:\ddot{O}-P-\ddot{O}: \quad :\ddot{O}: \right)^{-}$

2. $\ddot{S}=C=\ddot{S}$ ⟷ $:S\equiv C-\ddot{S}:$ ⟷ $:\ddot{S}-C\equiv S:$

3. $\left(:\ddot{O}-\dot{C}l-\ddot{O}: \quad :\ddot{O}: \right)^{-}$

4. $\ddot{O}=C=\ddot{S}$ linear; 180°; polar; sp; 2σ; 2π

5. $\left(\ddot{O}=N-\ddot{O}: \quad :\ddot{O}: \right)^{-}$ triangular planar; 120°; nonpolar; sp²; 3σ; 1π

6. $H-\overset{\overset{\displaystyle H}{|}}{\underset{\underset{\displaystyle H}{|}}{C}}-\ddot{O}-\overset{\overset{\displaystyle :\ddot{C}l:}{|}}{\underset{\underset{\displaystyle :\ddot{C}l:}{|}}{C}}-\ddot{C}l:$ bent; 109°; polar; sp³; 8σ

7. $\left(:\ddot{F}-\overset{\overset{\displaystyle :\ddot{F}:}{|}}{\underset{\underset{\displaystyle :\ddot{F}: \; \ddot{F}:}{|}}{Sb}}-\ddot{F}: \right)^{2-}$ square pyramid; sp³d²

8.

$$
\begin{array}{c}
\ddot{:}\ddot{F}\ddot{:} \\
| \\
:\ddot{C}l - P - \ddot{F}: \\
| \quad \diagdown \\
:\ddot{C}l: \quad :\ddot{C}l:
\end{array}
$$

trigonal bipyramid; sp³d

Thermochemistry

I. Principles of Heat Flow

  A. Basic vocabulary definitions

    1. System: _____

    _____

    2. Surroundings: _____

    _____

    3. State of a system: _____

    _____

    4. State property: _____

    _____

  B. Sign of heat flow

    1. The symbol for heat flow is _____ .

    2. q is + when _____ .

    3. q is – when _____ .

    4. Endothermic process

      a. Sign of q: _____

      b. Heat flows from _____ to _____ .

      c. Effect on the temperature of the surroundings: _____

      _____

    d. Example of an endothermic process: _____

_____

  5. Exothermic process

    a. Sign of q: _____

    b. Heat flows from _____ to _____.

    c. Effect on the temperature of the surroundings: _____

_____

    d. Example of an exothermic process: _____

_____

C. Magnitude of heat flow

  1. Mathematical equation that defines the magnitude of heat flow

  2. Heat capacity

    a. Symbol: _____

    b. Definition: _____

_____

    c. Equation for the heat capacity of a pure substance

    d. Units for heat capacity _____

  3. Specific heat

    a. Symbol: _____

    b. Definition: _____

_____

    c. Mathematical relationship between the magnitude of heat flow and specific heat

    d. Units for specific heat _____

e. *Exercises*

(1) The specific heat of ethyl alcohol is 2.43 J/g·°C. If the ethyl alcohol absorbs 17.85 J of heat when its temperature increases from 23.13°C to 27.48°C, what is the mass of the ethyl alcohol present?

Solution: This is a simple plug-in problem. We use the equation that relates the magnitude of heat flow, q, to specific heat.

$$q = c \times m \times \Delta t$$

where

$$c = 2.43 \; \frac{J}{g \cdot ^\circ C}$$

$$q = 17.85 \; J$$

$$\Delta t = t_{final} - t_{initial} = 27.48^\circ C - 23.13^\circ C = 4.35^\circ C$$

Note that the sign for q is positive because heat flows into the system (The problem says ethyl alcohol absorbs heat.) from the surroundings. Thus,

$$17.85 \; J = 2.43 \; \frac{J}{g \cdot ^\circ C} \times m \times 4.35^\circ C$$

$$m = \frac{17.85 \; J}{2.43 \; \frac{J}{g \cdot ^\circ C} \times 4.35^\circ C} = 1.69 \; g$$

(2) Calculate the specific heat of an alloy if 17.98 J of heat must be absorbed to raise the temperature of a 6.00-g sample from 20.00°C to 28.35°C. What is the heat capacity of the alloy? **(E1)**

II. Measurement of Heat Flow; Calorimetry
A. Coffee cup calorimeter
1. Formula for calculating heat flow into water

2. Relationship used to calculate q for the reaction

3. *Exercises*

    a. When 1.507 g of NH_4Cl are dissolved in 100.00 g of water at 25.00°C, the temperature drops to 23.98°C. Calculate q for the water and q for the solution process.

Solution: We determine q for water by using the relationship between specific heat of water and q.

$$q_{water} = c_{water} \times m_{water} \times (t_{final} - t_{initial})$$

$$= 4.184 \, \frac{J}{g \cdot °C} \times 100.00 \, g \times (23.98 - 25.00)°C$$

$$= -427 \, J$$

To determine q for the solution process, we plug in to the relation

$$q_{reaction} = -q_{water}$$

$$= -(-427 \, J) = 427 \, J$$

The reaction is endothermic.

    b. When 12.84 grams of $MgSO_4$ are dissolved in 250.00 g of H_2O at 23.76°C, 9.745 kJ of heat are evolved

    (1) What is q for the solution process?

    (2) What is q for the water?

    (3) What is the final temperature of the solution? **(E2)**

    Remember: Specific heat is in J/g·°C; hence all energy units, including q, must be in joules and not in kilojoules.

B. Bomb calorimeter

    1. Heat relationship formula:

$$q_{reaction} = -q_{cal}$$

2. *Exercises:*

   a. A bomb calorimeter has heat capacity 5.105 kJ/°C. The combustion of 1.000 g of ethanol, C_2H_5OH, in the bomb increases the temperature by 5.252°C. What is the heat flow associated with the combustion of the ethanol? What would $q_{reaction}$ be if one mol of ethanol was used instead? *Solution:* Again, this is a simple problem. Since

$$q_{reaction} = -q_{cal} = -(C_{cal} \times \Delta t)$$

   we substitute 5.105 kJ/°C and 5.252°C for C_{cal} and Δt, respectively. We obtain

$$q_{reaction} = -(5.105 \, \frac{kJ}{°C} \times 5.252°C) = -26.81 \, kJ$$

   This is the heat flow associated with the combustion of 1.000 g of ethanol. To obtain $q_{reaction}$ for one mole of ethanol, we use the fact that one mole of ethanol, C_2H_5OH, has a mass of 46.07 g/mol. Therefore, $q_{reaction}$ for one mole will be 46.07 times that for one gram. Hence

$$q_{reaction} = -26.81 \, \frac{kJ}{g} \times 46.07 \, \frac{g}{mol} = -1235 \, \frac{kJ}{mol}$$

   b. Calculate the heat capacity of a bomb calorimeter if 1.000 g of methyl alcohol, CH_3OH, is burned in oxygen and the reaction evolves 19.93 kJ of heat. The temperature of the bomb rises from 23.72°C to 27.10°C. **(E3)**

III. Enthalpy
A. Definition

_____

B. Symbol: _____

C. Relation to $q_{reaction}$:

D. ΔH
   1. Relationship of ΔH to enthalpy of products and reactants:

2. Sign of ΔH

    a. Exothermic reaction: _____

    b. Endothermic reaction: _____

E. Enthalpy is a state property.

F. *Exercises*

    1. When a piece of sodium metal is added to a beaker with water, the temperature in the beaker goes from 25°C to 42°C. Is the reaction exothermic or endothermic? What is the sign of ΔH?

    Solution: The reaction is exothermic, since the temperature of the surroundings increased. The sign of ΔH is negative.

    2. When ammonium nitrate is dissolved in water, the temperature drops.

       a. Is heat evolved to the surroundings or absorbed from the surroundings?

       b. Write the chemical equation for the reaction involved. **(E4)**

IV. Thermochemical Equations

A. Things to remember about a thermochemical equation:

    1. The sign of ΔH indicates whether the reaction, when *carried out at constant pressure,* is exothermic or endothermic.

    2. The coefficients of the balanced equation represent the number of moles of the species in that particular equation.

    3. The phases (physical states) of all species must be specified in the balanced equation.

       a. Symbol for liquid: _____

       b. Symbol for gas: _____

       c. Symbol for solid: _____

       d. Symbol for aqueous solution: _____

    4. The value of ΔH applies when products and reactants are at the same temperature. If the temperature is not specified, the reaction is assumed to have been carried out at _____.

B. Laws of Thermochemistry

    1. The magnitude of ΔH is directly proportional to the amount of reactant or product.

       a. The ΔH written to the right of the equation indicates how many kilojoules of energy are either evolved or absorbed by the reaction.

b. Once again, conversion factors can be obtained from the reaction. With a thermochemical equation, in addition to the mole and gram relationships you can write, you can also obtain a relationship involving moles, grams, and kilojoules for the substances in the reaction.

c. Consider the reaction between sodium metal and liquid water. It produces solid sodium hydroxide, hydrogen gas, and 281.8 kJ of heat energy.

(1) Write the thermochemical equation for this reaction.

$$2\,Na\,(s) + 2\,H_2O\,(\ell) \;\rightarrow\; 2\,NaOH\,(s) + H_2\,(g) \qquad \Delta H = -281.8\,kJ$$

Since heat is produced, the reaction is exothermic and $\Delta H < 0$.

(2) Write all the relationships you can obtain from this reaction.
2 mol Na = 45.98 g Na = 2 mol H_2O = 36.04 g H_2O = 2 mol NaOH = 80.00 g NaOH = 1 mol H_2 = 2.02 g H_2 = -281.8 kJ

(3) Write the conversion factor for converting between grams of water and kilojoules of energy.

$$\frac{36.04\,g\,H_2O}{-281.8\,kJ} \qquad or \qquad \frac{-281.8\,kJ}{36.04\,g\,H_2O}$$

d. *Remember*

(1) The sign of ΔH is included in every conversion factor that you write. Do *not* conveniently drop it.

(2) You will need *balanced* equations to proceed.

e. *Exercises*

(1) When four moles of iron metal are burned in oxygen, iron(III) oxide is formed and 1644.4 kJ of energy are evolved.

(a) Write a balanced thermochemical equation for this reaction, using whole number coefficients.

$$4\,Fe\,(s) + 3\,O_2\,(g) \;\rightarrow\; 2\,Fe_2O_3\,(s) \qquad \Delta H = -1644.4\,kJ$$

(b) How much heat is evolved if 3.031 g of iron are burned?
Solution: The equivalences for this reaction are
4 mol Fe = 223.4 g Fe = 3 mol O_2 = 96.00 g O_2 = 2 mol Fe_2O_3 = 319.4 g Fe_2O_3 = -1644.4 kJ.
We want to convert grams of Fe to kilojoules; hence, the conversion factor is

$$\frac{223.4\,g\,Fe}{-1644.4\,kJ} \qquad or \qquad \frac{-1644.4\,kJ}{223.4\,g\,Fe}$$

$$3.031\,g\,Fe \times \frac{-1644.4\,kJ}{223.4\,g\,Fe} = -22.31\,kJ$$

Thus 22.31 kJ of energy are evolved.

(2) When a mole of solid calcium sulfate reacts with carbon dioxide, solid calcium carbonate and sulfur trioxide gas are formed and 223.8 kJ of heat are absorbed.

(a) Write a balanced thermochemical equation for the reaction.

(b) How many grams of $CaCO_3$ can be produced if 4.843 kJ are absorbed? **(E5)**

2. ΔH for a reaction is equal in magnitude but opposite in sign to the ΔH of the reverse reaction.

a. Note that the reverse reaction must be exactly the same as the forward reaction, except for the direction of the arrowhead. Thus

$$2\,Ag\,(s) + Br_2\,(\ell) \rightarrow 2\,AgBr\,(s)$$

is the reverse of

$$2\,AgBr\,(s) \rightarrow 2\,Ag\,(s) + Br_2\,(\ell)$$

but not of

$$AgBr\,(s) \rightarrow Ag\,(s) + \frac{1}{2}\,Br_2\,(\ell)$$

Therefore, since

$$2\,Ag\,(s) + Br_2\,(\ell) \rightarrow 2\,AgBr\,(s) \qquad \Delta H = -199\,kJ$$

then,

$$2\,AgBr\,(s) \rightarrow 2\,Ag\,(s) + Br_2\,(\ell) \qquad \Delta H = +199\,kJ$$

To obtain the thermochemical equation for

$$AgBr\,(s) \rightarrow Ag\,(s) + \frac{1}{2}\,Br_2\,(\ell)$$

we use the conversion factor

$$\frac{199\,kJ}{2\;mol\;AgBr}$$

so,

$$1 \text{ mol AgBr} \times \frac{199 \text{ kJ}}{2 \text{ mol AgBr}} = 99.5 \text{ kJ}$$

and we can then write

$$\text{AgBr (s)} \rightarrow \text{Ag (s)} + \frac{1}{2} \text{ Br}_2 \text{ } (\ell) \qquad \Delta H = 99.5 \text{ kJ}$$

b. *Exercises*

When 1 mol of hydrogen gas reacts with 1 mol of iodine gas to produce hydrogen iodide gas, 51.8 kJ of energy are absorbed.

(1) Write a balanced thermochemical equation for this reaction

(2) How much heat is involved in the decomposition of 8.164 g of HI into H_2 and I_2?

(3) Is the heat in the decomposition absorbed or evolved? **(E6)**

3. The value of ΔH for a reaction is the same whether it occurs directly or in a series of steps. If Equation (c) is the sum of Equations (a) and (b), then ΔH for Equation (c) is the sum of the ΔH's for Equations (a) and (b). This is known as Hess's Law.

a. Most exercises using this law will give two equations and then ask you to determine ΔH for a third equation. The natural tendency is to assume that adding both equations as written will automatically give the third equation. This, however, is not always the case. Follow these simple steps to ensure that the two equations you add do give the actual overall equation.

(1) Check that the reactants in the overall equation are present as reactants in the equations being added. If not present as reactants, then reverse the equation(s) to achieve this. (Do not forget to reverse the sign of ΔH too!) The same holds true for the products.

(2) Check that the compounds that are not in the overall reaction but are in the equations being added cancel each other out; that is, they appear as a reactant in one reaction and as a product in the other. If this is not the case, rearrange the equations to achieve this. Do *not* rearrange or do anything to the overall equation.

(3) Check that the reactants and products have the same number of moles in both the overall reaction and the equations that are being added. If this is not the case, multiply or divide the equation by an integer to achieve this result.

Caution:
(a) Do not do this to the overall equation.
(b) Do not forget to include ΔH when you are multiplying or dividing the entire equation by an integer.

b. Exercises

(1) Add the following thermochemical equations

$$2\,Na_2O_2\,(s) + 2\,H_2O\,(\ell) \rightarrow$$
$$4\,NaOH\,(s) + O_2\,(g) \qquad \Delta H = -126.4\ kJ \quad (1)$$

$$NaOH\,(s) + HCl\,(g) \rightarrow$$
$$NaCl\,(s) + H_2O\,(\ell) \qquad \Delta H = -179.2\ kJ \quad (2)$$

to produce the overall reaction

$$2\,Na_2O_2\,(s) + 4\,HCl\,(g) \rightarrow 4\,NaCl\,(s) + O_2\,(g) + 2\,H_2O\,(\ell)$$

Solution: Checks:
(a) Reactants in the overall reaction are: Na_2O_2 and HCl.
Both are reactants in the composite equations (Equations (1) and (2)).
Products in the overall equation are: NaCl, H_2O, and O_2. All are products in the composite equations.
(b) Compound(s) not present in overall equation but present in composite equations: NaOH.
NaOH appears as a product in Equation (1) and a reactant in Equation (2). The directions of Equations (1) and (2) are all right.
(c) Number of moles:
Na_2O_2: 2 mol in overall equation, 2 mol in Equation (1) — okay.
HCl: 4 mol in overall equation, 1 mol in Equation (2). Thus, multiply Equation (2) by 4 to get 4 mol of HCl there, too.

$$4\,NaOH\,(s) + 4\,HCl\,(g) \rightarrow$$
$$4\,NaCl\,(s) + 4\,H_2O\,(\ell) \qquad \Delta H = 4(-179.2)\ kJ$$

H_2O: 2 mol in overall equation, 4 mol of product in revised Equation (2) and 2 mol of reactant in Equation (1), thus, net H_2O is 2 mol of product — okay.

$$2\,Na_2O_2\,(s) + 2\,H_2O\,(\ell) \rightarrow$$
$$4\,NaOH\,(s) + O_2\,(g) \qquad \Delta H = -126.4\ kJ \quad (1)$$

$$4\,NaOH\,(s) + 4\,HCl\,(g) \rightarrow$$
$$4\,NaCl\,(s) + 4\,H_2O\,(\ell) \qquad \Delta H = -716.8\ kJ \quad (2)$$

(d) The overall equation therefore is

$$2\,Na_2O_2\,(s) + 4\,HCl\,(g) \longrightarrow$$
$$4\,NaCl\,(s) + O_2\,(g) + 2\,H_2O\,(\ell) \qquad \Delta H = -843.2\ kJ$$

This procedure looks long and cumbersome, but with practice you will build self-confidence and soon be able to do it by inspection.

(2) Use the equations

$$2\,C_2H_2\,(g) + 5\,O_2\,(g) \longrightarrow$$
$$4\,CO_2\,(g) + 2\,H_2O\,(g) \qquad \Delta H = -2512\ kJ$$

$$N_2\,(g) + \frac{1}{2}\,O_2\,(g) \longrightarrow N_2O\,(g) \qquad \Delta H = 104\ kJ$$

to obtain ΔH for the reaction

$$C_2H_2\,(g) + 5\,N_2O\,(g) \longrightarrow 2\,CO_2\,(g) + H_2O\,(g) + 5\,N_2\,(g) \qquad \textbf{(E7)}$$

c. Hess's law can also be used to determine ΔH for phase changes when both phase and temperature change. Remember, you cannot change both the temperature and the phase in the same equation. You have to keep the phase constant and change the temperature, and then keep the temperature constant and change the phase.

Exercises

(1) Calculate ΔH for the process

$$H_2O\,(\ell,\ 75.0°C) \longrightarrow H_2O\,(g,\ 102.0°C)$$

Solution: This overall reaction consists of the following steps:

$$H_2O\,(\ell,\ 75.0°C) \longrightarrow H_2O\,(\ell,\ 100.0°C) \qquad \Delta H_1$$

Here, we changed the temperature and kept the phase constant. We did not go from 75.0°C to 102.0°C in the liquid state, because at 1 atm, water can exist as a liquid only till 100°C. The next step is

$$H_2O\,(\ell,\ 100.0°C) \longrightarrow H_2O\,(g,\ 100.0°C) \qquad \Delta H_2$$

Notice that in this step, we kept the temperature constant at 100°C and changed the state. The third and final step is

$$H_2O \, (g \, , \, 100.0°C) \; \rightarrow \; H_2O \, (g, \; 102.0°C) \qquad \Delta H_3$$

We now calculate all the ΔH's of the intermediate steps:
For ΔH_1 where we only changed the temperature, we get

$$\Delta H_1 = q_{reaction} = c_{water} \times m \times \Delta t$$

$$= 4.184 \, \frac{J}{g \cdot °C} \times 1 \, mol \, H_2O \times 18.0 \, \frac{g}{mol} \times 25.0°C$$

$$= 1.88 \times 10^3 \, J = 1.88 \, kJ$$

For ΔH_2, we see from Table 8.2 that ΔH_{vap} for one mole of H_2O is 40.7 kJ.
ΔH_3 is calculated in a similar manner to ΔH_1.

$$\Delta H_3 = q_{reaction} = 4.184 \times 18.0 \times (102.0 - 100.0) = 0.15 \, kJ$$

Using Hess's law, we now add ΔH_1, ΔH_2, and ΔH_3 to get ΔH for the overall equation.

$$\Delta H = 1.88 + 40.7 + 0.15 = 42.7 \, kJ$$

(2) Calculate ΔH for the following equation
$$H_2O \, (\ell \, , \, 50.0°C) \; \rightarrow \; H_2O \, (s, \; -10.0°C) \qquad \textbf{(E8)}$$

C. Using calorimetry data to write thermochemical equations

Knowing the amount of heat evolved or absorbed by a reaction gives a conversion factor that you can use to calculate ΔH for the equation. Consider the exercise with the coffee cup calorimeter. Here you were given data to calculate ΔH for the solution of 1.507 g of NH_4Cl. You calculated ΔH to be 427 J. Your conversion factor is therefore

$$\frac{1.507\ g}{427\ J} \quad \text{or} \quad \frac{427\ J}{1.507\ g}$$

The reaction is

$$NH_4Cl\ (s)\ \rightarrow\ NH_4^+\ (aq) + Cl^-\ (aq)$$

You therefore want ΔH for the dissolving of one mole of NH_4Cl. You thus want to convert moles of NH_4Cl to kilojoules. Since you do not have a conversion factor involving moles and kilojoules directly, you will have to convert 1 mole of NH_4Cl to grams of NH_4Cl using the conversion factor 53.49 g of NH_4Cl = 1 mole. Now you can convert 53.49 g of NH_4Cl to kilojoules.

$$53.49\ g\ NH_4Cl \times \frac{427\ J}{1.507\ g} \times \frac{1\ kJ}{1000\ J} = 15.2\ kJ$$

The thermochemical equation is therefore

$$NH_4Cl\ (s)\ \rightarrow\ NH_4^+\ (aq) + Cl^-\ (aq) \qquad \Delta H = 15.2\ kJ$$

Exercises

1. Using the data from the combustion of ethanol (C_2H_5OH) in the bomb calorimeter (page 145), write the thermochemical equation if the combustion of ethanol in the bomb involves the reaction between ethanol and oxygen gas to produce carbon dioxide and steam. **(E9)**

2. Using the data from the combustion of methyl alcohol (CH_3OH) in the bomb calorimeter (E3), write the thermochemical equation if the burning of methyl alcohol in oxygen in the bomb produces carbon dioxide gas and steam. **(E10)**

V. Enthalpies of Formation

A. Definition: _____

_____

B. Formula for determining $\Delta H°$

C. Heat of formation of elements in their stable state is

_____.

D. Formula for determining $\Delta H°$ for ions in water solution

E. Heat of formation of H^+: _____

F. *Exercises*

    1. The reaction between one mole of sodium oxide and water yields an aqueous solution of sodium ions and hydroxide ions. The reaction evolves 237.5 kJ of heat.

       a. Write a balanced thermochemical equation using smallest single whole number coefficients.

       Solution:

$$Na_2O\,(s) + H_2O\,(\ell) \rightarrow 2\,Na^+\,(aq) + 2\,OH^-\,(aq) \qquad \Delta H° = -237.5\,kJ$$

       Remember: Since you are told that 237.5 kJ are evolved, ΔH must be negative.

      b. Calculate the heat of formation of sodium oxide using Table 8.3 in your textbook.

       Solution:

$$\Delta H° = \sum \Delta H_f° \text{ products} - \sum \Delta H_f° \text{ reactants}$$

$$= [2\,\Delta H_f°(Na^+) + 2\,\Delta H_f°(OH^-)] - [\Delta H_f°(Na_2O) + \Delta H_f°(H_2O)]$$

$$-237.5\,kJ = \left[\left(2\,mol \times -240.1\,\frac{kJ}{mol}\right) + \left(2\,mol \times -230.0\,\frac{kJ}{mol}\right)\right] -$$

$$\left[(\Delta H_f°\,Na_2O) + \left(1\,mol \times -285.8\,\frac{kJ}{mol}\right)\right]$$

$$= (-480.2\,kJ - 460.0\,kJ) - (\Delta H_f°\,Na_2O - 285.8\,kJ)$$

$$\Delta H_f°\,Na_2O = -416.9\,\frac{kJ}{mol}$$

Note that the units for $\Delta H°$ are different from the units for $\Delta H_f°$. The unit for $\Delta H°$ is kJ, whereas that for $\Delta H_f°$ is kJ/mol.

2. The balanced thermochemical equation for the reaction between sulfur dioxide gas and liquid bromine is

$$Br_2\,(\ell) + SO_2\,(g) + 2\,H_2O\,(\ell) \;\longrightarrow$$
$$4\,H^+\,(aq) + 2\,Br^-\,(aq) + SO_4^{2-}\,(aq) \qquad \Delta H° = -280.9\,kJ$$

Calculate the heat of formation of the sulfur dioxide gas. **(E11)**

VI. Bond Enthalpy

A. Definition: _____

_____

B. Compare the bond enthalpy of single bonds to that of multiple bonds:

_____

_____

C. Estimation of ΔH from bond enthalpies
   1. The enthalpy change, ΔH, of a reaction can be determined from bond enthalpies only if all the reactants and products are gases.
   2. Bond breaking requires energy. Suppose you are interested in the H—Br bond. Refer to Table 8.4 of your text. Look up H in the vertical row of elements and Br in the horizontal row of elements. The point at which these two lines meet is the bond enthalpy involved with the bond. Breaking that bond requires energy; so, ΔH for that process is +368 kJ. Making a bond liberates energy; thus, forming the H—Br bond liberates 368 kJ, or ΔH for the bond-making process is –368 kJ.
   3. Note that the bond enthalpy for a single bond between two atoms is different from the bond enthalpy of a double or triple bond between the same two atoms.
   4. The bond enthalpies of multiple bonds are larger than those of single bonds. The number of π bonds present increases the bond enthalpy between the two atoms. The bond enthalpy of a double bond, however, is not equal to the sum of two single bond enthalpies.

5. *Exercises*

    a. What is the enthalpy change in the formation of the N—O bond? **(E12)**

    b. What is the enthalpy change in the breaking up of a C=C bond? **(E13)**

VII. First Law of Thermodynamics

  A. Types of energy in thermodynamics

    1. Heat

      a. Symbol: _____

      b. Positive when: _____

      c. Negative when: _____

    2. Work

      a. Symbol: _____

      b. Positive when: _____

      c. Negative when: _____

  B. First law of thermodynamics

    1. Statement: _____

    _____

    2. Mathematical statement:

    3. *Exercises*

      a. Calculate ΔE for a gas involved in a process for which the gas evolves 25 J of heat and does 48 J of work by expanding.
      Solution:

$$w = -48 \text{ since work is done on the surroundings}$$
$$q = -25 \text{ since heat is evolved}$$
$$\Delta E = q + w$$
$$= -25 + (-48) = -73 \text{ J}$$

b. A gas has ΔE of -115 J. The gas absorbs 38 J of heat. How much work is involved in the process? Is it done by the system or to the system? **(E14)**

C. ΔH versus ΔE
   1. Constant volume
      a. A reaction that takes place at constant volume (e.g. bomb calorimeter) with no mechanical or electrical work involved has $w = 0$.
      b. The heat flow (q_v) for this type of reaction is equal to _____ .
   2. Constant pressure
      a. A reaction that takes place in an open container is said to be done at constant pressure.
      b. Here heat flow (q_p) is equal to _____ .
   3. Mathematical relationship between ΔH and ΔE
      a. Mathematical relationship

$$\Delta H = \Delta E + \Delta (PV)$$

      b. The unit for PV is L-atm, while the unit for both ΔH and ΔE is kJ. The relationship between L-atm and kJ is _____ .
      c. $\Delta(PV) = (PV)_{products} - (PV)_{reactants}$
         $\Delta(PV) = $ (moles of gaseous products – moles of gaseous reactants)PV
         For

$$2H_2\,(g) + O_2\,(g) \rightarrow 2H_2O\,(\ell)$$

$$\Delta(PV) = [0 - (2 + 1)]PV = -3PV$$

      d. For 1 mole of gas at 25°C, PV = 24.5 L-atm.

$\Delta E \approx$ _____

      e. Values for ΔH and ΔE are generally close to one another because ΔPV is ordinarily a small quantity.

SELF-TEST
A. Multiple Choice:

1. The heat of formation of liquid water refers to the equation
   a. $2 H_2 (g) + O_2 (g) \rightarrow 2 H_2O (\ell)$
   b. $2 H (g) + O (g) \rightarrow H_2O (\ell)$
   c. $H^+ (aq) + OH^- (aq) \rightarrow H_2O (\ell)$
   d. $H_2 (\ell) + \frac{1}{2} O_2 (g) \rightarrow H_2O (\ell)$
   e. None of these

2. When the heat content of the reactants in a reaction is greater than the heat content of the products
   a. heat is absorbed and the reaction is exothermic.
   b. heat is evolved and the reaction is exothermic.
   c. heat is absorbed and the reaction is endothermic.
   d. heat is evolved and the reaction is endothermic.
   e. none of the above applies.

3. When 2.0 g of a certain solid are added to 50.0 g of water in a calorimeter, the temperature drops from 25.0 to 22.6°C. This means that
   a. q_{H_2O} is positive and the reaction is endothermic.
   b. q_{H_2O} is positive and the reaction is exothermic.
   c. q_{H_2O} is negative and the reaction is endothermic.
   d. q_{H_2O} is negative and the reaction is exothermic.

4. Given that

   $$2 Fe_2O_3 (s) \rightarrow 4 Fe (s) + 3 O_2 (g) \qquad \Delta H° = +393.0 \, kJ$$

   then $\Delta H_f°$ for Fe_2O_3 is

   a. +196.5 kJ   b. −196.5 kJ   c. +393.0 kJ   d. −393.0 kJ

5. Which of the following data is not required for calculating ΔH of solution for one mole of sample in a coffee cup calorimeter?
   a. temperature change of the water   b. mass of the water
   c. atmospheric pressure   d. mass of the sample
   e. specific heat of the water

6. In a reaction, $\Delta H = +312$ kJ. How many of the following statements are true?

   — The products have a lower enthalpy than the reactants.
   — The enthalpy change for the reverse reaction is −312 kJ.
   — The reaction absorbs heat from the surroundings.
   — The reverse reaction is endothermic.

   a. 0   b. 1   c. 2   d. 3   e. 4

7. Which of the following are true about ΔH of a reaction?

   (1) ΔH for the overall reaction does not depend on the number of steps required to complete the reaction.
   (2) ΔH is related to the heats of formation of the reactants and products.
   (3) ΔH is independent of the amounts of products or reactants.
   (4) ΔH has the same sign regardless of the direction of the reaction.

   a. (2)   **b.** (1),(2)   **c.** (2),(3)   **d.** (2),(4)   **e.** (1),(2),(3),(4)

8. The bond energy for the H—Cl bond is equal to
   a. ΔH when one mole of H—Cl bonds is formed from gaseous atoms.
   b. the heat of formation of H—Cl.
   c. ΔH for the process $H(g) + Cl(g) \rightarrow HCl(g)$.
   d. ΔH for the process $H_2(g) + Cl_2(g) \rightarrow 2\,HCl(g)$.
   e. none of these.

9. The standard heat of formation of HF is -268.6 kJ/mol. Which of the following reactions represents $-$ B.E. for HF? (B.E. stands for bond energy)
   a. $H(g) + F(g) \rightarrow HF(g)$
   b. $2H(g) + 2F(g) \rightarrow 2\,HF(g)$
   c. $\frac{1}{2}H_2(g) + \frac{1}{2}F_2(g) \rightarrow HF(g)$
   d. $H_2(g) + F_2(g) \rightarrow 2\,HF(g)$

10. Given $\Delta H° = -635.1$ kJ for the reaction

$$Ca(s) + \frac{1}{2}O_2(g) \rightarrow CaO(s)$$

What would you expect to happen if the reaction were allowed to proceed at 1 atm in a heavily insulated container so that no heat flow could occur between the reaction mixture and the surroundings?
   a. No reaction would occur.
   b. The temperature of the reaction mixture would rise.
   c. The temperature of the reaction mixture would fall.
   d. Cannot tell because there is not enough information given.

B. Answer the questions below, using **LT** (for *less than*), **GT** (for *greater than*), **EQ** (for *equal to*), or **MI** (for *more information required* in the blanks provided.

_____ **1.** The energy required to break a C=C bond __(1)__ the energy required to break a C—C bond.

_____ **2.** Consider two metals A and B ($c_A > c_B$) of equal mass and at the same temperature. Consider further two beakers X and Y each with 100 g of water at 25°C. Metal A is added to beaker X, while metal B is added to beaker Y. The temperature of the water in both beakers rises. The temperature of the water in beaker X __(2)__ the water temperature in beaker Y.

_____ **3.** In an endothermic reaction, $q_{reactants}$ _(3)_ $q_{products}$.

_____ **4.** For the reaction

$$H_2 + O_2 \rightarrow H_2O$$

$\Delta H°$ is _(4)_ to the heat of formation of H_2O (ℓ).

_____ **5.** The heat of solution for compound Z is -100 kJ/mol. When Z is dissolved in water at 25°C, the temperature of the resulting solution is _(3)_ 25°C.

C. True or False

Consider a reaction with the following energy diagram.

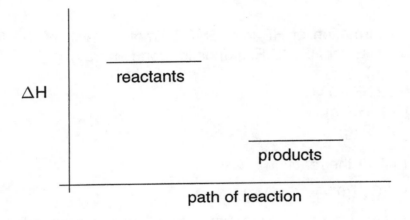

For this reaction

_____ **1.** the temperature of the surroundings increases.

_____ **2.** $q_{reaction} > 0$ at constant pressure.

_____ **3.** it is an exothermic reaction.

_____ **4.** a change in the mass of reactants changes the sign for ΔH.

_____ **5.** the enthalpy of products is larger than the enthalpy of reactants.

D. Problems
Consider the reaction between one mole of solid silver sulfide ($\Delta H_f^\circ = -31.8$ kJ/mol) and liquid water, producing silver metal, hydrogen sulfide (H_2S) gas, and oxygen gas.
  1. Write the thermochemical equation for this reaction using Table 8.3 of your text.

  2. Is the reaction exothermic or endothermic?

  3. How much heat is required to produce 10.00 g of silver?

  4. If only 200.0 kJ of heat are supplied and 400.0 g of silver sulfide are used as a starting reagent (an excess of water is available), how many grams of silver sulfide will remain unreacted?

Consider the reaction between two moles of lithium metal and water to produce solid lithium hydroxide and hydrogen gas.

5. Write an equation for the reaction.

6. If the reaction is exothermic and evolves 402.4 kJ of heat, write the thermochemical equation.

7. Calculate ΔH_f° for LiOH.

8. Calculate the final temperature of the water if the reaction was done in a bomb calorimeter (heat capacity = 6.03×10^3 J/°C) at 25.00°C.

ANSWERS

Exercises:

(E1) specific heat = 0.359 J/g·°C; heat capacity = 2.15 J/°C

(E2) $q_{reaction}$ = −9.745 kJ; q_{H_2O} = 9.745 kJ; t_{final} = 33.08°C

(E3) 5.90 kJ/°C

(E4) NH_4NO_3 (s) $\rightarrow$ NH_4^+ (aq) + NO_3^- (aq) $\Delta H > 0$

(E5) a. $CaSO_4$ (s) + CO_2 (g) $\rightarrow$ SO_3 (g) + $CaCO_3$ (s) $\Delta H°$ = 223.8 kJ; **b.** 2.166 g

(E6) (1) I_2 (g) + H_2 (g) $\rightarrow$ 2 HI (g) $\Delta H°$ = 51.8 kJ;   **(2)** −1.65 kJ;   **(3)** evolved

(E7) $\Delta H°$ = −1776 kJ

(E8) −10.52 kJ

(E9) C_2H_5OH (ℓ) + 3 O_2 (g) $\rightarrow$ 2 CO_2 (g) + 3 H_2O (g) $\Delta H°$ = −1235 kJ

(E10) 2 CH_3OH (ℓ) + 3 O_2 (g) $\rightarrow$ 2 CO_2 (g) + 4 H_2O (g) $\Delta H°$ = −1277 kJ

(E11) −300.0 kJ/mol   **(E12)** −222 kJ   **(E13)** 612 J

(E14) −153 kJ, done by the system

Self-Test

A. Multiple Choice:

1. e	**2.** b	**3.** c	**4.** b	**5.** c
6. c	**7.** b	**8.** e	**9.** a	**10.** b

B. Greater than, Less than, or Equal to:

1. GT	**2.** GT	**3.** LT	**4.** MI	**5.** LT

C. True or False:

1. T	**2.** F	**3.** T	**4.** F	**5.** F

D. Problems:

1. Ag_2S (s) + H_2O (ℓ) $\rightarrow$ 2 Ag (s) + H_2S (g) + $\frac{1}{2}$ O_2 (g) $\Delta H°$ = 297.0 kJ

2. endothermic

3. 13.76 kJ

4. 233.1 g

5. 2 Li (s) + 2 H_2O (ℓ) $\rightarrow$ 2 LiOH (s) + H_2 (g)

6. 2 Li (s) + 2 H_2O (ℓ) $\rightarrow$ 2 LiOH (s) + H_2 (g) $\Delta H°$ = −402.4 kJ

7. −487.0 kJ

8. 91.7 °C

9

Liquids and Solids

I. **Liquid-Vapor Equilibrium**
  A. Dynamic equilibrium

  1. Definition: _____

  _____

  2. Once equilibrium is reached, the concentration of the molecules in the vapor does not change with time.
  B. Vapor pressure

  1. Definition: _____

  _____

  2. Vapor pressure vs. volume of the container
    a. As long as liquid and vapor are both present, the pressure exerted by the vapor is independent of the volume of the container. Once all the liquid is gone, the pressure is dependent on the volume of the container according to the ideal gas law.
    b. *Remember:*
      (1) The pressure in the container cannot exceed the vapor pressure at the given temperature.
      (2) The pressure in the container is either equal to the vapor pressure (in which case both liquid and vapor are present) or <u>less</u> than the vapor pressure (in which case only gas is present).
      (3) Only the vapor obeys the ideal gas law.
    c. To determine whether all the liquid in the container will evaporate at a given volume, determine the mass of the vapor that the container will hold. Use the ideal gas law for your calculations, taking V to be the volume of the container, P to be the vapor pressure of the liquid, and T to be the temperature in the container.

d. *Exercises*

(1) A 50.0-L tank at 25°C contains 25.0 g of acetone (C_3H_6O). The vapor pressure of acetone at that temperature is 229.2 mm Hg. Is liquid acetone present in the tank? If not, how much acetone should be in the tank so that some liquid remains?

Solution: To answer the first question, we use the ideal gas law to calculate how many grams of acetone at a pressure of 229.2 mm Hg will be present in the container as a vapor. We use the formula

$$m = \frac{PV(MM)}{RT}$$

Substituting the values given, we get

$$m = \frac{\frac{229.2}{760} \times 50.0 \times 58.1}{0.0821 \times 298} = 35.8 \, g$$

This means that, at a pressure of 229.2 mm Hg, the container should have 35.8 g of acetone vapor. Since there are only 25.0 g of acetone in the tank, it does not contain any liquid acetone. Indeed, the pressure in the tank is not even 229.2 mm Hg. Can you figure out what it is? (Answer: 160 mm Hg)

The second question asks how many grams of acetone are needed to have some liquid in the tank. Earlier, we calculated that 35.8 g of acetone were required just to maintain a pressure of 229.2 mm Hg. Hence any mass of acetone *above* 35.8 g will remain as a liquid in equilibrium with the vapor.

(2) At 25.0°C, a 25.0-L tank of bromine gas contains 55.0 g of bromine. The vapor pressure of bromine gas at 25.0°C is 214 mm Hg. A tank of chlorine gas at the same temperature and volume has a pressure of 5.0 atm. The vapor pressure of chlorine gas at the same temperature is 7.69 atm. Describe the physical states of the halogens present in each of the tanks. How many grams of gas are present in each tank? **(E1)**

C. Vapor pressure vs. temperature
   1. Clausius-Clapeyron equation

   2. *Exercises*
      a. Calculate the temperature at which the vapor pressure of water is 1000 mm Hg (3 significant figures), given that ΔH_{vap} is 4.07×10^4 J/mol at 100.0°C.
      Solution: It is usually a good idea when you solve problems using the Clausius-Clapeyron equation to identify as (1) and (2) the temperatures and their corresponding pressures. In this case, we let $T_1 = 100.0$°C. Since this is the boiling point of water, P_1 will be 760 mm Hg. T_2 is unknown and P_2 is 1000 mm Hg. Pressure units can be either in mm Hg or atmospheres as long as they are consistent. R is not the gas constant 0.0821 L·atm/mol·K. It should be in energy units (8.31 J/mol·K). Make sure that ΔH_{vap} is in joules. it is most often given in kJ.
      Substituting into the Clausius-Clapeyron equation, we get

$$\ln 1000 - \ln 760 = \frac{4.07 \times 10^4 \, \dfrac{J}{mol}}{8.31 \, \dfrac{J}{mol \cdot K}} \left(\frac{1}{373} - \frac{1}{T_2} \right)$$

Working out the left side of the equation, we get

$$6.908 - 6.633 = 0.275$$

If you do not know how to use your calculator to take logarithms, or you do not remember the algebraic rules for logarithms, you should get help now. You will see this type of equation again in later chapters.
We now work the right side (the side with the unknown) of the equation.

$$\frac{4.07 \times 10^4}{8.31} = 4.90 \times 10^3$$

and

$$\frac{1}{373} = 0.00268$$

Putting our left and right sides together, we get the simplified equation

$$0.275 = 4.90 \times 10^3 \left(0.00268 - \frac{1}{T_2} \right)$$

Simplifying further

$$0.275 = 13.1 - \frac{4.90 \times 10^3}{T_2}$$

Rearranging, we get

$$\frac{4.90 \times 10^3}{T_2} = 13.1 - 0.275$$

Finally, solving for T_2 we get

$$T_2 = \frac{4.90 \times 10^3}{13.1 - 0.275} = 381 \text{ K}$$

The chemistry of problems involving the Clausius-Clapeyron equation is not particularly difficult. It is the algebra that often confuses many students. If the problem asks for a pressure, then you should work out the right side of the equation first. After you have finished the calculations of the right side, take INV LN of both sides. You have to take the inverse ln of the right side on your calculator too. Once you have done this, the rest is easy.

b. Calculate the vapor pressure of water at 61.0°C if the vapor pressure of water at 70.0°C is 234 mm Hg and its heat of vaporization is 4.07×10^4 J/mol. **(E2)**

3. Experimental determination of heat of vaporization

$$\Delta H_{vap} \left(\frac{J}{mol} \right) = -(2.30)(8.31) \times slope$$

D. Boiling point
  1. Definition: _____

  _____

  2. Normal boiling point
  Definition: _____

  _____

E. Critical temperature and pressure
   1. Definition of critical temperature: _____

   _____

   2. Definition of critical pressure: _____

   _____

II. Phase Diagrams
   A. The diagram
      1. Definitions:
         a. Phase diagram: _____

         _____

         b. Triple point: _____

         _____

   2. *Exercises*
      Consider the phase diagram for CO_2.

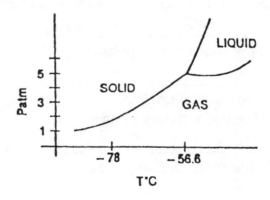

   a. At what temperature and pressure does the triple point occur?
      Solution: Looking at the graph, we see that the triple point is the point at
      which all three branches intersect. In the case of CO_2, the triple point is 5
      atm and −56.6°C.
   b. Describe the phase changes, if any, that occur when a sample of CO_2 at
      −80°C and 0.1 atm is compressed at constant temperature to a pressure of
      7 atm. **(E3)**

    c. Describe the phase changes, if any, that occur when a sample of CO_2 at −80°C and 0.1 atm is heated at a constant pressure to a temperature of −50°C. **(E4)**

B. Sublimation

   1. Definition: _____

      _____

   2. A solid sublimes at any temperature below the triple point when the pressure above it is reduced below the equilibrium vapor pressure.

   3. A solid melts at a temperature above the triple point.

C. Melting and freezing points

   1. Definition: _____

      _____

   2. Relation of melting point to the triple point of most substances:

      _____

   3. Effect of pressure on the more dense phase:

      _____

   4. Effect on the solid-liquid equilibrium line

     a. When the solid is the more dense phase:

      _____

     b. When the liquid is the more dense phase:

      _____

III. Molecular Substances and Intermolecular Forces

  A. Properties of molecules

   1. Electrical conductivity: _____

      _____

   2. Solubility in water: _____

      _____

   3. Melting and boiling points: _____

      _____

  B. The force or bond that holds the atoms (which make up the molecule) together

    is _____.

    This force is called intramolecular (within the molecule) force.

C. Intermolecular forces

   1. They are the forces between the molecules, electrical in nature, and involve an attraction between temporary or induced forces.

   2. Dispersion (London) forces

      a. They are found in all molecular substances, both polar and nonpolar.

      b. They are the only forces between nonpolar molecules.

      c. They increase in strength with increasing molar mass.

      d. Effect of dispersion forces on boiling point

        _____

   3. Dipole forces

      a. Remember, dipole forces are intermolecular forces that arise as a result of the interaction between adjacent polar molecules. They are not to be confused with the polar covalent bonds that hold several atoms together to make a molecule. Dipole forces align themselves very similarly to ionic forces except that they are weaker.

      b. Effect of dipole forces on melting and boiling points

        _____

   4. Hydrogen bonds

      a. What is a hydrogen bond?

        _____

        _____

      b. A hydrogen bond exists only between a hydrogen atom attached to a fluorine, oxygen, or nitrogen atom and the fluorine, oxygen, or nitrogen atom of another molecule.

        Why are hydrogen bonds stronger than dipole forces?

        (1) _____

        _____

        _____

        (2) _____

        _____

      c. Properties of water resulting from the presence of hydrogen bonds:

        (1) _____

        _____

        (2) _____

        _____

5. *Exercises*
   a. What type of forces would you expect in the following molecules?

   H_2, H_2S, $H—CF_3$

   Solution: Hydrogen (H_2) has only dispersion forces since it is a nonpolar molecule.

   H_2S has dipole forces as well as dispersion forces.

   $H—CF_3$ has dipole forces as well as dispersion forces. This molecule does not have hydrogen bonds because the H atom is not attached to an F, O, or N atom. Instead, it is attached to a carbon atom.

   b. What type of forces would you expect in CO_2, HF, and BrCl? **(E5)**

IV. Network covalent, Ionic, and Metallic Solids

   A. Network covalent solids

   1. The atoms are held together by _____ .

   2. Properties of network covalent solids

   a. Melting points: _____
   _____

   b. Electrical conductivity: _____
   _____

   c. Solubility in water: _____
   _____

   3. Two network covalent carbon solids
   a. Diamond
   (1) Hybridization: _____
   (2) Geometry: _____
   (3) Diamond is one of the hardest substances known.
   b. Graphite
   (1) Hybridization: _____
   (2) Geometry: _____
   4. Quartz
   a. Formula: _____
   b. Silicon bonds to _____ atoms of oxygen.
   c. Geometry is _____ .

B. Ionic solids
    1. Two structural units for ionic solids:

        a. _____

        b. _____

    2. Most often, compounds formed by _____ are ionic.
    3. Properties of ionic solids

        a. Melting points: _____

        _____

        b. Electrical conductivity: _____

        _____

        c. Solubility in water: _____

        _____

C. Metals
    1. Metals are anchored in position by _____

    _____.

    2. Properties of metals

        a. Melting points: _____

        _____

        b. Electrical and thermal conductivity: _____

        _____

        c. Solubility in water: _____

        _____

        d. Ductility
        Definition: _____

        e. Malleability
        Definition: _____

        f. Luster
        Definition: _____

V. Crystal Structures
    A. Unit cell definition:

_____

_____

B. Simple cubic cell (SC)

   1. Description of the unit cell

      a. Location of atoms that make up the unit cell:

_____

      b. Atoms touch at _____ .

      c. Number of atoms per unit cell: _____

   2. Relationship between the atomic radius (r) and the length of one side of the cell (s):

C. Face-centered cubic cell (FCC)

   1. Description of unit cell

      a. Location of atoms that make up the unit cell:

_____

      b. Atoms touch at _____ .

      c. Number of atoms per unit cell: _____

   2. Relationship between the atomic radius (r) and the length of one side of the cell (s):

D. Body-centered cubic cell (BCC)

   1. Description of unit cell

      a. Location of atoms that make up the unit cell:

_____

      b. Atoms touch at _____ .

      c. Number of atoms per unit cell: _____

   2. Relationship between the atomic radius (r) and the length of one side of the cell (s):

E. *Exercises*

   1. Gold crystallizes in a face-centered cubic structure and has an atomic radius of 0.144 nm. What is the length of a side of the unit cell for the gold crystal? What is the volume of the unit cell?

     Solution: For a face-centered cubic cell, the relationship between the atomic radius and the length of a side is

$$r = \frac{s\sqrt{2}}{4}$$

Substituting the values given in the exercise into the equation, we get

$$s = \frac{4r}{\sqrt{2}} = \frac{4 \times 0.144}{\sqrt{2}} = 0.407 \, \text{nm}$$

Since the volume of the cell is equal to the cube of its side, the volume of the cell is $(0.407 \text{ nm})^3 = 0.0674 \text{ nm}^3$.

2. Silver crystallizes in a body-centered cubic structure (BCC) with a side length of 0.41 nm. What is the atomic radius of a silver atom? What is the volume of a unit cell? If the mass of the unit cell is 7.2×10^{-22} g, what is the density of the cell in g/cm^3? **(E6)**

SELF-TEST
A. Multiple Choice:

1. The critical temperature of methane is −82°C. This means that methane
   a. boils at −82°C.
   b. can never be a liquid at room temperature.
   c. can never be a liquid at −95°C.
   d. contains strong hydrogen bonds.

2. Which of the following shows hydrogen bonding?
   a. CH_4 **b.** H_2 **c.** H_2O
   d. all of these   **e.** none of these

3. For a molecular substance, which of the following is NOT true?
   a. The heat of vaporization (ΔH_{vap}) is a measure of the strength of the intermolecular forces in a liquid.
   b. The lower the melting point of a solid, the weaker are the intermolecular forces in general.
   c. ΔH_{vap} may be obtained from a plot of ln P vs 1/T.
   d. The critical point defines the temperature above which a gas may not be liquefied, no matter how great the pressure.
   e. The pressure of the vapor above a liquid depends upon the volume of the container.

4. A solid melts at 75°C. It is insoluble in water and does not conduct electricity in solution, as a solid, or in the molten state. The compound is most likely
   a. metallic   **b.** ionic   **c.** molecular   **d.** network covalent

5. Which of the following statements is true?
  a. C_2H_5OH contains hydrogen bonds but water does not.
  b. Ionic compounds generally have higher melting points than molecular compounds, because ionic bonds are weak compared to the intermolecular bonds in molecular compounds.
  c. Since Ne has larger dispersion forces than He, neon has a higher boiling point.
  d. All of the above statements are true.
  e. None of the above statements are true.

6. The boiling point of carbon tetrachloride, CCl_4, is 77°C, while that of chloroform, $CHCl_3$, is 61°C. This is because
  a. the gravitational attraction between CCl_4 molecules is greater than that of $CHCl_3$.
  b. CCl_4 is polar and $CHCl_3$ is not.
  c. $CHCl_3$ has hydrogen bonding.
  d. the dispersion forces in CCl_4 are greater than the combined dispersion forces in $CHCl_3$.

7. A solid is in equilibrium with its vapor at a certain temperature and pressure. As the volume of the system is decreased while maintaining equilibrium, which of the following is(are) true?

  (1) The pressure goes down.
  (2) The pressure goes up.
  (3) Solid sublimes.
  (4) Gas is converted to solid.

  a. (2),(3)   **b.** (1),(3)   **c.** (3)   **d.** (4)   **e.** (2),(4)

8. Which of the following is(are) true?

  (1) critical temperature < triple point temperature
  (2) critical temperature > triple point temperature
  (3) critical pressure < triple point pressure
  (4) critical pressure > triple point pressure
  (5) critical temperature = triple point temperature

  a. (3)   **b.** (2),(3),(5)   **c.** (2),(4)   **d.** (1),(3)   **e.** (1),(2),(5)

9. How many of the following groups correctly list the compounds in order of *increasing* boiling point?

  (1) $CH_3CH_2CH_2CH_3 < CH_3CH_2CH_2OH < CH_3CH_2OCH_3 < OHCH_2CH_2OH$
  (2) $H_2S < H_2Se < H_2Te < H_2O$
  (3) $CH_4 < CH_3CH_3 < CH_3CH_2CH_3 < CH_3CH_2CH_2CH_3$

  a. 0   **b.** 1   **c.** 2   **d.** 3

10. How many of the following are molecular substances?
      Na_4SiO_4 SiO_2 Fe I_2 H_2O C (graphite)
  a. 1   **b.** 2   **c.** 3   **d.** 4   **e.** 5   **f.** 6

B. Fill in the blanks:

Write the compound with the *higher* boiling point in the blanks provided.

_____ 1. C_2H_5OH or C_4H_{10}

_____ 2. H_2O or H_2S

_____ 3. Br_2 or ICl (iodine chloride)

_____ 4. $NaCl$ or SCl_2

_____ 5. C_3H_8 or $C_{12}H_{26}$

_____ 6. HF or HCl

C. True or False:

1. KNO_3, an ionic solid

_____ a. does not conduct electricity.

_____ b. is soluble in water.

_____ c. can conduct electricity when the solid is melted.

_____ d. has covalent bonds in its structure.

_____ e. has an equal number of cations and anions when dissolved in water.

2. The boiling point of ammonia (NH_3) is $-33°C$, while the boiling point of phosphine (PH_3) is $-87.7°C$. Ammonia has a higher boiling point because

_____ a. the molar mass of NH_3 is less than that of PH_3.

_____ b. dispersion forces for NH_3 are stronger than those for PH_3.

_____ c. NH_3 is hydrogen bonded while PH_3 is not.

_____ d. NH_3 is polar while PH_3 is not.

_____ e. NH_3 is ionic while PH_3 is not.

3. The normal boiling point of a liquid

_____ a. is the only temperature at which liquid and vapor are in equilibrium.

_____ b. is the temperature at which the vapor pressure is 1.00 atm.

_____ c. varies with atmospheric pressure.

_____ d. is the temperature at which the vapor pressure equals external pressure.

_____ e. can be reduced by increasing the pressure.

D. Less than, greater than, Euqal to:

Answer the questions below, using **LT** (for *is less than*), **GT** (for *is greater than*), **EQ** (for *is equal to*), or **MI** (for *more information required*) in the blanks provided.

_____ 1. At 50°C, benzene has a vapor pressure of 269 mm Hg. A flask that contains both benzene liquid and vapor at 50°C has a pressure   (1)   269 mm Hg

_____ 2. Ether has a vapor pressure of 537 mm Hg at 25°C. A flask that contains only ether vapor at 37°C has a pressure   (1)   537 mm Hg.

_____ 3. The boiling point of H_2O_2 (1)   the boiling point of C_3H_8.

_____ 4. The energy required to vaporize liquid bromine   (1)   the energy required to decompose Br_2 into Br atoms.

_____ 5. The dispersion forces present in naphthalene, $C_{10}H_8$,   (1)   the dispersion forces present in butane, C_4H_{10}.

E. Problems:

1. A certain element crystallizes in a body-centered cubic cell. It has an atomic radius of 0.186 nm.
   a. What is the length of one side of the cell?

   b. What is the volume of the unit cell in cm^3?

2. The vapor pressure of water is 17.5 mm Hg at 20°C. Calculate the heat of vaporization of water.

3. Substance X has a vapor pressure of 175 mm Hg at 25°C.
   a. Substance X is placed in a 15-L tank with a pressure gauge that registers 100 mm Hg at 25°C. What phase(s) of substance X is(are) in the tank?

   b. Substance X is transferred from the 15-L tank to a 30-L tank at the same temperature with no loss of the substance. What phase(s) is(are) present in the tank at equilibrium? Support your answer with calculations.

4. Given the following hypothetical phase diagram:

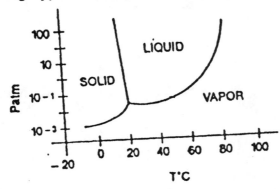

   a. Which phase (solid or liquid) of this substance is the denser phase?

   b. Describe any phase changes that would occur if the following sequence of steps was followed:
   At constant temperature, the vapor at 60°C and 0.001 atm is compressed to 100 atm and then cooled to −100°C. Keeping the temperature at −100°C, the pressure is gradually reduced until it is 10^{-4} atm.

    c. Describe any phase changes that would occur if solid is gradually heated at a constant pressure of 0.01 atm from a temperature of 10°C to 60°C.

ANSWERS
Exercises:
(E1) Br_2 tank: liquid and gas, 46.0 g gas; Cl_2 tank: only gas, 3.6×10^2 g gas

(E2) 159 mm Hg

(E3) It goes from gas to solid.

(E4) no change

(E5) CO_2 — dispersion forces; HF — H–bond, dipole and dispersion forces
      BrCl — dipole and dispersion forces.

(E6) r = 0.18 nm; V = 0.069 nm^3; density = 1.0×10 g/cm^3

Self-Test
A. Multiple Choice:

1. b	**2.** c	**3.** e	**4.** c	**5.** c
6. d	**7.** d	**8.** c	**9.** c	**10.** b

B. Fill in the Blanks:

1. C_2H_5OH	**2.** H_2O	**3.** ICl	**4.** NaCl	**5.** $C_{12}H_{26}$
6. HF				

C. True or False:

1. a) T	**b)** T	**c)** T	**d)** T	**e)** T
2. a) F	**b)** F	**c)** T	**d)** F	**e)** F
3. a) F	**b)** T	**c)** F	**d)** F	**e)** F

D. Less than, Greater than, Equal to:

1. EQ	**2.** MI	**3.** GT	**4.** LT	**5.** GT

E. Problems:

    1. a. 0.430 nm   **b.** 7.95 $\times 10^{-23}$ cm^3

    2. 4.28 $\times 10^4$ J/mol

    3. a. all gas   **b.** all gas

    4. a. liquid   **b.** vapor to liquid to solid to vapor   **c.** solid to vapor

10
Solutions

I. Concentration Units
A. Molarity

   1. Symbol for molarity: _____

   2. Formula for molarity:

   3. What a given molarity can tell you

      a. The number of moles of solute in a liter of solution

      Example: A 0.250 M solution of Na_2SO_4 has 0.250 mol of Na_2SO_4 in one liter of solution. Remember solution is not the same as solvent. A solution that is 0.250 M does *not* mean 0.250 mol of Na_2SO_4 in one liter of water.

      b. The number of moles of solute in a given volume of solution

      You can use the molarity of a solution as a conversion factor, with which you can convert a given volume of solution to the number of moles of solute in that solution.

      Example: How many moles of Na_2SO_4 are there in 0.300 L of solution that is 0.250 M?

$$0.250\,M \quad \text{means} \quad \frac{0.250\,\text{mol } Na_2SO_4}{1\,\text{L solution}}$$

Thus,

$$0.300\,L \;\times\; \frac{0.250\,\text{mol } Na_2SO_4}{1\,L} = 0.0750\,\text{mol}$$

      c. The volume of solution that is necessary, given the number of moles of solute and the molarity of the solution

Example: What is the volume of a 0.250 M solution that has 0.700 mol of Na_2SO_4 as solute?

Again, we use the conversion factor

$$\frac{0.250 \text{ mol } Na_2SO_4}{1 \text{ L solution}}$$

to convert 0.700 mol of solute to a volume of solution. Thus,

$$0.700 \text{ mol } \times \frac{1 \text{ L solution}}{0.250 \text{ mol } Na_2SO_4} = 2.80 \text{ L}$$

d. The number of grams of solute in a liter of solution

Example: How many grams of Na_2SO_4 are in a liter of a solution that is 0.250 M?

Here you have to use the conversion factor

$$\frac{142 \text{ g } Na_2SO_4}{1 \text{ mol } Na_2SO_4}$$

Thus, since 0.250 M means 0.250 mol Na_2SO_4/L of solution, we get

$$\frac{0.250 \text{ mol } Na_2SO_4}{\text{L solution}} \times \frac{142 \text{ g } Na_2SO_4}{1 \text{ mol } Na_2SO_4} = \frac{35.5 \text{ g } Na_2SO_4}{\text{L solution}}$$

e. The number of grams of solute in a given volume of solution

Example: How many grams of Na_2SO_4 are in 0.800 L of a 0.250 M solution?

Again the conversion factors are:

$$\frac{142 \text{ g } Na_2SO_4}{1 \text{ mol } Na_2SO_4} \qquad \text{and} \qquad \frac{0.250 \text{ mol } Na_2SO_4}{\text{L solution}}$$

Here 0.800 L has to be converted to grams of Na_2SO_4. So

$$0.800 \text{ L} \times \frac{0.250 \text{ mol } Na_2SO_4}{\text{L solution}} \times \frac{142 \text{ g } Na_2SO_4}{1 \text{ mol } Na_2SO_4} = 28.4 \text{ g } Na_2SO_4$$

f. The volume of a solution with a given molarity required for a mass of solute

Example: What is the volume of a solution that has 157 g of Na_2SO_4 if the solution is 0.250 M?

The conversion factors again are:

$$\frac{142 \text{ g } Na_2SO_4}{1 \text{ mol } Na_2SO_4} \qquad \text{and} \qquad \frac{0.250 \text{ mol } Na_2SO_4}{\text{L solution}}$$

This time we have to change 157 g Na_2SO_4 to liters of solution. Thus,

$$157 \text{ g } Na_2SO_4 \times \frac{1 \text{ mol } Na_2SO_4}{142 \text{ g } Na_2SO_4} \times \frac{\text{L solution}}{0.250 \text{ mol } Na_2SO_4} = 4.42 \text{ L solution}$$

4. *Exercises*

   a. Calculate the molarity of a solution prepared by adding enough water to 73.2 g of ammonium nitrate to make 0.835 L of solution. **(E1)**

   b. How many moles of calcium chloride do you need in 356 mL of solution to make a 0.125 M solution? **(E2)**

   c. How many grams of potassium permanganate are required in 565 mL of solution to obtain a 0.380 M solution? **(E3)**

   d. What volume of solution is required for a 1.25 M solution of $Pb(NO_3)_2$ that contains 58.5 g of solute? **(E4)**

e. How many moles of phosphate and ammonium ions are there in 0.250 L
of a 0.150 M solution of ammonium phosphate? **(E5)**

Caution: It can be very confusing to do molarity problems using ratios and
proportions. The factor conversion method is simpler and clearer.

5. Preparing solutions by dilution
   a. Formula used for dilution problems:

   b. *Exercises*
   (1) How would you prepare 100.0 mL of a 0.200 M solution of NaOH from
   a bottle that is labeled 0.4871 M?

   Solution: The concentrated solution is the one that is on hand. In
   this case it is the one labeled 0.4871 M. The dilute solution is the one
   that has to be prepared. In this case it is 0.200 M. What you want to
   find is the volume of concentrated solution to take so that you can add
   enough water to make 100.0 mL of a 0.200 M solution. Thus, V_c is
   unknown, M_c is 0.4871 M, V_d is 100.0 mL, and M_d is 0.200 M. Using
   the formula, we get

   $$V_c = \frac{M_d \times V_d}{M_c} = \frac{0.200\,M \times 100.0\,mL}{0.4871\,M} = 41.1\,mL$$

   You would prepare the final solution by taking 41.1 mL of the initial
   solution and adding enough water to make the final volume 100.0 mL.
   Note that in dilution problems, the volumes of the solutions do not have
   to be in liters. You just have to be consistent. In calculating molarity for
   other problems, however, you must always have the volume in liters.
   (2) Calculate the molarity of a solution prepared by taking 25.00 mL of a
   0.400 M solution of HNO_3 and adding enough water to make 75.00
   mL of solution. **(E6)**

B. Mole fraction

Recall from Chapter 5 that the mole fraction of a component in a mixture is given by the relation

$$X_A = \frac{\text{number of moles of A}}{\text{total number of moles of all components}}$$

C. Mass percent

1. Formula for determining mass percent:

2. *Exercises*

a. Calculate the mass percent of water in a solution made up of 123 g of NaOH and 289 g of water.

Solution: Since

$$\text{mass percent of water} = \frac{\text{mass of water}}{\text{mass of solution}} \times 100$$

we get

$$\text{mass percent of water} = \frac{289\,\text{g}}{123\,\text{g} + 289\,\text{g}} \times 100 = 70.1\,\%$$

b. An aqueous solution of potassium iodide is often made up by mixing KI and I_2 and dissolving the mixture in water. A 5.06% KI solution has 18 g of KI and I_2 and 35 g of water. How much KI is in the mixture? **(E7)**

D. Parts per million (ppm)

1. Formula for determining parts per million:

2. Mathematical relation between parts per million and mass percent:

3. *Exercises*

   a. Methyl mercury(II) chloride, CH_3HgCl, is a toxic pollutant usually found in waters around industrial sites. If a sample of 250.0 mL of solution ($d = 1.00$ g/mL) is found to contain 0.200 mg of CH_3HgCl, what is the concentration of methyl mercury(II) chloride in parts per million?

   Solution: The formula we use is

   $$ppm = \frac{mass\ of\ solute}{total\ mass\ of\ solution} \times 10^6$$

   The mass of the solute, CH_3HgCl, is 0.200 mg, which is 2.00×10^{-4} g. The mass of the solution should also be in grams. Since we are given a volume and density, we obtain

   $$250.0\ mL\ of\ solution \times \frac{1.00\ g\ solution}{1.00\ mL\ solution} = 2.50 \times 10^2\ g\ solution$$

   Substituting now into the formula for ppm, we get

   $$ppm = \frac{2.00 \times 10^{-4}\ g}{2.50 \times 10^2\ g} \times 10^6 = 0.800$$

   b. A water solution contains 0.56 ppm of lead as a pollutant. If you drank a cup (1/2 pint) of this water, how many mg of lead would you ingest? (density of water = 1.00 g/mL) **(E8)**

E. Molality

   1. Symbol for molality: _____

   Do not confuse the symbol for molality (m) with the symbol for mass (m) that we used earlier. The symbol for molality is always in italics.

   2. Formula for determining molality:

   3. *Exercises*

   a. A solution contains sucrose, $C_{12}H_{22}O_{11}$, in water. The solution has a total mass of 738 g and the mass percent of sucrose in the solution is 38.4%. Calculate the molality of the solution.

   Solution: To calculate the molality of the solution, you need to know two things: the mass of the solute and the mass of the solvent. You can calculate the mass of the solute, sucrose, by using the information on the mass percent of the solution.

$$38.4 = \frac{\text{mass of sucrose}}{738 \text{ g solution}} \times 100$$

Solving for the mass of sucrose, we get

$$m_{\text{sucrose}} = \frac{38.4 \times 738}{100} = 283 \text{ g}$$

Since the solution consists of water and sucrose, the mass of water is

$$738 \text{ g} - 283 \text{ g} = 455 \text{ g}$$

Since molality is moles of solute/kilogram of solvent, we need to convert grams of solute to moles of solute, and grams of solvent to kilograms of solvent. Starting with the solute, we need the molar mass of sucrose if we are to calculate moles from mass. The molar mass of sucrose is

$$12(12.01) + 22(1.01) + 11(16.0) = 342 \text{ g}$$

Using this molar mass, we now convert to moles.

$$283 \text{ g sucrose} \times \frac{1 \text{ mol sucrose}}{342 \text{ g sucrose}} = 0.827 \text{ mol}$$

The unit for the solvent has to be changed from grams to kilograms.

$$455 \text{ g} \times \frac{1 \text{ kg}}{1000 \text{ g}} = 0.455 \text{ kg}$$

The molality is therefore

$$m = \frac{0.827 \text{ mol solute}}{0.455 \text{ kg}} = 1.82$$

b. A 0.633 m solution of vitamin B$_6$ (pyridoxine) in water is prepared by adding 53.5 g of vitamin B$_6$ to 5.00×10^2 g of water. Calculate the molar mass of vitamin B$_6$. **(E9)**

F. Relating concentration units to each other

1. If you examine all concentration units closely, you see that the numerator of each unit is either in grams of solute or in moles of solute. The denominator in each unit is the one that varies. When you want to convert from one concentration unit to another, it is wise to separate the numerator from the denominator and work with them as separate entities. After you have gotten the required units for both numerator and denominator, put them together.

2. Given a concentration unit, it is also a good idea to choose one unit arbitrarily for the denominator. For example, if you were asked to calculate the molarity of a 2.50 molal solution, then choose one kilogram as the unit for the solvent since that is the denominator unit of the given molality. The numerator is of course 2.50 moles of solute, which is also the numerator for molarity.

3. It is also important to know the density in most of these interconversions.

4. You will find that organizing your data by filling in a prepared table is the easiest way to do these interconversions. You may not need all the entries, but you will find that the data is much more accessible arranged this way. A blank table should look like this:

	mass (g)	moles	volume (L)
solute			X
solvent			X
solution			

We put X's for the volumes of solute and solvent because we will never need those entries. The use of the table will be shown in the following exercise.

5. *Exercises*

   a. What is the molarity of an aqueous solution of HCl that is 35.0% by mass and whose density is 1.19 g/mL?

   Solution: Since we are given mass percent (mass of solute/100 g of solution), we arbitrarily choose to work with 100.0 grams of solution. Doing that, we can deduce that there are 35.0 g of solute and 65.0 g of solvent. Your table should now look like this:

	mass (g)	moles	volume (L)
solute	35.0		X
solvent	65.0		X
solution	100.0		

Knowing that the solute is HCl (MM = 36.46 g/mol) and that the solvent is water, (MM = 18.02 g/mol), we can fill in the table with the results of the following calculations.

$$35.0 \text{ g HCl} \times \frac{1 \text{ mol}}{36.46 \text{ g}} = 0.960 \text{ mol HCl}$$

$$65.0 \text{ g H}_2\text{O} \times \frac{1 \text{ mol}}{18.02 \text{ g}} = 3.61 \text{ mol H}_2\text{O}$$

$$\text{mol solution} = 0.960 + 3.61 = 4.57$$

	mass (g)	moles	volume (L)
solute	35.0	0.960	X
solvent	65.0	3.61	X
solution	100.0	4.57	

The last blank in the table, volume of solution, can be obtained by using the density of the solution to convert mass of the solution to volume.

$$100.0 \text{ g solution} \times \frac{1.000 \text{ mL solution}}{1.19 \text{ g solution}} \times \frac{1 \text{ L}}{1000 \text{ mL}} = 0.0840 \text{ L}$$

Your table should now look like this:

	mass (g)	moles	volume (mL)
solute	35.0	0.960	X
solvent	65.0	3.61	X
solution	100.0	4.57	0.0840

Note that we did all this without worrying about what was asked for. Now, no matter what concentration unit is asked for, you can easily calculate it, provided that you know the definition for that concentration unit. In this case, the unit asked for is molarity. Thus using the definition for molarity and picking out the appropriate data from the table, we get

$$M = \frac{\text{mol solute}}{\text{L solution}} = \frac{0.960 \text{ mol HCl}}{0.0840 \text{ L solution}} = 11.4$$

We could, however, just as easily have solved for molality or mole fraction of HCl.

b. How would you prepare 5.00 L of 2.00 M H_2SO_4 from a bottle labeled: H_2SO_4 — 98.0% by mass; density — 1.84 g/mL? **(E10)**

c. Calculate the molality of a 2.500 M solution of sulfuric acid. Its density is 1.200 g/mL.

   Solution: We choose 1.000 L of solution, which therefore means that we have 2.500 moles of solute (from the definition of molarity). We list these as our initial entries in the table.

	mass (g)	moles	volume (mL)
solute		2.500	X
solvent			X
solution			1.000

There are many ways in which you can now proceed. We choose to determine the mass of the solution by using the density of the solution that is given.

$$1000 \text{ mL solution} \times \frac{1.200 \text{ g solution}}{1 \text{ mL solution}} = 1.200 \times 10^3 \text{ g solution}$$

Our table now looks like this:

	mass (g)	moles	volume (mL)
solute		2.500	X
solvent			X
solution	1.200×10^3		1.000

We now determine the mass of the solute using the molar mass of H_2SO_4.

$$2.500 \text{ mol } H_2SO_4 \times \frac{98.08 \text{ g}}{1 \text{ mol}} = 245.2 \text{ g } H_2SO_4$$

We enter that value in our table.

	mass (g)	moles	volume (mL)
solute	245.2	2.500	X
solvent			X
solution	1.200×10^3		1.000

Now, we can determine the mass of solvent ($1.200 \times 10^3 - 245.2$) and the number of moles of solvent, water. Finally, we determine the moles of solution by adding moles of solute and moles of solvent. The completed table now looks like this.

	mass (g)	moles	volume (mL)
solute	245.2	2.500	X
solvent	955	53.0	X
solution	1.200×10^3	55.5	1.000

Since the problem asks for the molality of the solution, we pick the moles of solute from the table (2.500 mol) and the mass of solvent in kg (0.955). Thus

$$m = \frac{2.500 \text{ mol solute}}{0.955 \text{ kg solvent}} = 2.62$$

d. Calculate the mole fraction of H_2SO_4, its mass percent, and parts per million for the solution in *c*. **(E11)**

II. Principles of Solubility
A. Solute-solvent interactions
1. Solubility is dependent on both the nature of the solute and the solvent.
2. Electrical forces that determine the extent to which ionic compounds dissolve in water are

   a. _____

   _____

   b. _____

   _____

3. "Like dissolves like" principle
   a. Ionic substances tend to dissolve in _____ solvents.
   b. Nonpolar or slightly polar solutes are most soluble in solvents that are

      (polarity) _____ .

B. Temperature effects on solubility
1. Solid solute in liquid solvent
   Ordinarily, dissolving a solid is an endothermic process. Thus, an increase in temperature most often favors the solubility of a solid solute in a liquid solvent.
2. Saturation
   a. Unsaturated solution
      One that has less solute than it is capable of dissolving at a particular temperature
   b. Saturated solution
      One that has the exact amount of solute that it is capable of dissolving at a particular temperature
   c. Supersaturated solution

      Definition: _____

      _____

3. Gas solute in liquid solvent
   Ordinarily, dissolving a gas in a liquid is an exothermic process. Therefore, a decrease in temperature favors the solubility of the gas in a liquid.

C. Pressure effects on solubility
1. Solid solute in a liquid solvent
   No pressure effects noticeable
2. Gas solute in liquid solvent
   An increase in pressure increases the solubility of a gas in a liquid.
3. Henry's Law
   Mathematical statement:

III. Colligative Properties of Solutions

A. Definition of colligative properties:

_____

_____

B. Colligative properties of nonelectrolytes
  1. Vapor pressure lowering
    a. Raoult's Law
      Mathematical expression:

    b. Exercises
      (1) The vapor pressure of pure ethanol, C_2H_5OH, at 20°C is 39.98 mm Hg. What is the vapor pressure of ethanol over a solution containing 5.0% by mass of iodine?
      Solution: To be able to use Raoult's Law, which says that the vapor pressure of the solvent over the solution is equal to the mole fraction of the solvent multiplied by the vapor pressure of the pure solvent, we must express the concentration of the solvent (ethanol) in terms of mole fraction.
      Assume 100 g of solution (since mass percent is given). The solution has 5.0% iodine and 95.0% ethanol. In grams, it has 5.0 g of iodine and 95.0 g of ethanol. The mole fraction of ethanol (X_1) is then

$$X_1 = \frac{\text{mol ethanol}}{\text{mol ethanol} + \text{mol I}_2} = \frac{\dfrac{95.0}{46.1}}{\dfrac{95.0}{46.1} + \dfrac{5.0}{253.8}} = 0.990$$

We now plug into Raoult's Law:

$$P_1 = 0.990 \times 39.98 = 39.6 \text{ mm Hg}$$

      (2) At 64°C the vapor pressure of pure benzene is 400.0 mm Hg. Calculate the vapor pressure of benzene above a solution prepared by mixing 25.0 g of benzene (C_6H_6) and 15.0 g of naphthalene ($C_{10}H_8$). What is the vapor pressure lowering? **(E12)**

2. Freezing point depression and boiling point elevation
  a. Mathematical expression for the determination of the freezing point depression of a solution:

  b. Mathematical expression for the determination of the boiling point elevation of a solution:

  c. *Exercises*
    (1) The freezing point of benzene is 5.50°C. Its freezing point constant is 5.12°C/m. How much p-dichlorobenzene ($C_6H_4Cl_2$) must be added to 50.0 g of benzene (C_6H_6) to lower the freezing point of the solution to that of pure water?
    Solution: The formula you need is

$$\Delta T_f = k_f \times m$$

ΔT_f means freezing point depression, which is the difference between the freezing point of the pure solvent (5.50°C) and that of the solution (0.00°C).

$$\Delta T_f = 5.50°C - 0.00°C = 5.50°C$$

k_f is given: 5.12°C/m. Hence, we must calculate the molality of the solution before we can determine the mass of p-dichlorobenzene (PCB).

$$m = \frac{\Delta T_f}{k_f} = \frac{5.50}{5.12} = 1.07$$

1.07 m means 1.07 mol PCB/kg benzene. We now obtain the mass of PCB in 50.0 g (0.0500 kg) benzene.

$$0.0500 \text{ kg } C_6H_6 \times \frac{1.07 \text{ mol PCB}}{1 \text{ kg } C_6H_6} \times \frac{147 \text{ g PCB}}{1 \text{ mol PCB}} = 7.86 \text{ g PCB}$$

(2) Methylene chloride (CH_2Cl_2) has a boiling point of 40.7°C and a boiling point constant of 2.49 °C/m. What is the boiling point of a 20.0 % (by mass) solution of ferrocene, ($Fe(C_5H_5)_2$), in methylene chloride? **(E13)**

3. Osmotic pressure

   a. Osmosis — definition: _____

   _____

   _____

   b. Symbol used for osmotic pressure: _____

   c. Mathematical relationship between osmotic pressure and molarity:

4. Molar masses of nonelectrolytes from colligative properties

Colligative properties are most often used to determine the molar masses of unknown nonelectrolyte solutes.

Exercises

   a. Calculate the molar mass of an unknown substance, if when 6.32 g of the unknown is added to 50.0 g of camphor, the solution formed has a boiling point of 212.00°C.

   Solution: You need to use the formula

$$\Delta T_b = k_b \times m$$

According to Table 10.2 of your text, k_b for camphor is 5.61°C/m. Since the boiling point of pure camphor is 207.42°C, the boiling point elevation is 4.58°C. We can therefore calculate the molality of the solution.

$$m = \frac{\Delta T_b}{k_b} = \frac{4.58}{5.61} = 0.816$$

Using the definition for molality we get

$$0.816 \, m = \frac{\text{moles of unknown solute}}{0.0500 \, \text{kg camphor}}$$

and

$$\text{moles of unknown solute} = 0.816 \times 0.0500 = 0.0408\,\text{mol}$$

We now find the molar mass by substituting into the definition for molar mass

$$\text{molar mass} = \frac{6.32\,\text{g}}{0.0408\,\text{mol}} = 155\,\frac{\text{g}}{\text{mol}}$$

b. A 0.0344-g sample of starch in 10.0 mL of water at 25°C has an osmotic pressure of 4.0 mm Hg. What is the approximate molar mass of the starch? **(E14)**

C. Colligative properties of electrolytes
   1. Colligative properties of dilute solutions are directly proportional to the concentration of solute particles. Because electrolytes in solution split into ions, there are more particles in solution in an electrolyte than there are in solutions of nonelectrolytes. For this reason, electrolytes have a greater effect on colligative properties than do nonelectrolytes.
   2. The equations for determining the colligative properties of electrolytes are the same as those for nonelectrolytes. The only difference is that in determining the freezing point depression, the boiling elevation, and the osmotic pressure for electrolytes, a multiplier i is added to the right hand side of the equation. The multiplier i stands for the number of particles in solution. For a nonelectrolyte, the multiplier is always one. For an electrolyte like NaCl, i = 2 because NaCl splits up into two ions. In determining the vapor pressure lowering for an electrolyte in solution, the difference is a bit more subtle. You will have to consider the number of moles of each ion in determining the mole fraction of the solute.

SELF-TEST
A. Multiple Choice:

1. Dried beans swell on soaking in water. This is due to
  a. deliquescence   **b.** supersaturation
  c. osmosis   **d.** solubility

2. The solubility of a gas in water generally
  a. decreases with increased pressure.
  b. increases with increased pressure.
  c. is independent of pressure.
  d. increases with increased temperature.
  e. is independent of temperature.

3. Consider a solution of 0.100 mol NaCl in 1.000 kg of water in a closed container at 75°C. When the solution is cooled, it contracts. How many of the following concentration variables change?

mass% of solute	mole fraction	molarity	molality

  a. 0   **b.** 1   **c.** 2   **d.** 3   **e.** 4

4. Concerning solubility, how many of the following statements are TRUE?

  — The solubility of butanol, C_4H_9OH, in water is greater than the solubility of butane, C_4H_{10}, in water.
  — Butanol, C_4H_9OH, is more soluble in hexane, C_6H_{14}, than is butane, C_4H_{10}.
  — The concentration of CO_2 in a bottle of soda decreases after the bottle is opened.
  — The oxygen available to fish in lakes is greater at the water's surface because the temperature there is higher.

  a. 0   **b.** 1   **c.** 2   **d.** 3   **e.** 4

5. Consider 0.100 m aqueous solutions of the following compounds.

  sugar   KI Na_2CO_3 $CaCl_2$ $CuSO_4$

  — The freezing point of the Na_2CO_3 solution is below all the others.
  — Another solution in the list above has the same freezing point as the Na_2CO_3 solution.
  — The KI and $CuSO_4$ solutions have the same freezing point.
  — The freezing point of the sugar solution is below all the others.

  The number of correct statements is
  a. 0   **b.** 1   **c.** 2   **d.** 3   **e.** 4

6. Which of the following correctly lists the compounds in order of *increasing* solubility in water?
  a. CCl_4, CH_2Cl_2, CH_3OH **b.** CCl_4, CH_3OH, CH_2Cl_2
  c. CH_2Cl_2, CH_3OH, CCl_4 **d.** CH_2Cl_2, CCl_4, CH_3OH
  e. CH_3OH, CCl_4, CH_2Cl_2

7. At 25°C, CO_2 dissolves in water with the release of heat. At the same temperature, KNO_3 dissolves in water with the absorption of heat. If the temperature is raised to 50°C and the pressure is kept constant, what happens to the solubilities of CO_2 and KNO_3?

   a. The solubility of CO_2 decreases while the solubility of KNO_3 decreases.
   b. The solubility of CO_2 increases while the solubility of KNO_3 decreases.
   c. The solubility of CO_2 decreases while the solubility of KNO_3 increases.
   d. The solubility of CO_2 increases while the solubility of KNO_3 increases.
   e. The solubility of CO_2 is unchanged while the solubility of KNO_3 increases.

B. Less than, Equal to, Greater than:

Consider the following beakers:

- Beaker A has 1.00 kg of a solvent Y and 0.100 mol of glucose ($C_6H_{12}O_6$).
- Beaker B has 1.00 kg of a solvent Y and 0.100 mol of $CaCl_2$.
- Beaker C has 1.00 kg of water and 0.100 mol of sucrose ($C_{12}H_{22}O_{11}$).

Answer the questions below, using **LT** (for *is less than*), **GT** (for *is greater than*), **EQ** (for *is equal to*), or **MI** (for *more information required*) in the blanks provided.

_____ **1.** The vapor pressure of pure water ___(1)___ the vapor pressure of the solution in Beaker C.

_____ **2.** The freezing point of the solution in Beaker A ___(2)___ the freezing point of the solution in Beaker B.

_____ **3.** The boiling point of the solution in Beaker A ___(3)___ the boiling point of the solution in Beaker C.

_____ **4.** The boiling point of the solution in Beaker B ___(4)___ the boiling point of the solution in Beaker C.

_____ **5.** The molality of the solution in Beaker A ___(5)___ the molality of the solution in Beaker C.

_____ **6.** The vapor pressure of the solution in Beaker A ___(6)___ the vapor pressure of the solution in Beaker C.

C. Fill in the blanks:

_____ 1. Which is more soluble in water? CH_3CH_2OH or $H_3C-O-CH_3$?

_____ 2. An increase in pressure would increase the solubility in water of NH_3 (g) or NaCl(s)?

_____ 3. Which of the 2 solids (AX or BY) would be more soluble if the temperature were increased?

        (1)   AX (s) $\rightarrow$ A^+ (aq) + X^- (aq) $\Delta H < 0$

        (2)   BY (s) $\rightarrow$ B^+ (aq) + Y^- (aq) $\Delta H > 0$

_____ 4. Of the following 4 concentration units (M, m, % by mass, ppm), which cannot be obtained by making only mass measurements?

_____ 5. If the pressure of CO_2 over water is tripled, then the solubility of CO_2 in water will be __(5)__ . (tripled, reduced by a third, increased 9-fold, reduced by a ninth)

D. True or False:

_____ 1. The solubility of CO_2 in water is less than the solubility of CH_3OH in water at 25°C.

_____ 2. The osmotic pressure of a solution at 100°C is equal to the osmotic pressure of the same solution at 0°C.

_____ 3. The boiling point of a pure solvent is greater than the boiling point of its solutions.

_____ 4. The freezing point depression of a 1 m aqueous solution of CH_3OH is less than that of a 1 m aqueous solution of NaCl.

_____ 5. The molality of an aqueous NaCl solution that freezes at -1°C is equal to the molality of another solution of NaCl that boils at 101°C.

_____ 6. The solubility of a gas with a Henry's Law constant of 0.0005 M/atm is less than that of a gas with a Henry's law constant of 0.002 M/atm at the same temperature.

E. Problems:

Consider a solution of aqueous $(NH_4)_2CO_3$ prepared by mixing 12.00 g of $(NH_4)_2CO_3$ with 125 g of water (density = 1.00 g/mL) to make 127 mL of solution.

1. Calculate the percent by mass of $(NH_4)_2CO_3$.

2. Calculate the mole fraction of $(NH_4)_2CO_3$.

3. Calculate the molarity of the solution.

4. Calculate the molality of the solution.

Consider an aqueous solution of potassium hydroxide.

5. How many liters of 3.50% KOH by mass with density 1.012 g/mL can be obtained by diluting 0.250 L of 30.0% KOH by mass (density = 1.288 g/mL)?

Consider an aqueous solution of nitric acid.

6. Calculate the molarity, mass percent, and mole fraction of nitric acid in an 8.92 m solution. The density of the solution is 1.22 g/mL.

Consider oxygen.

7. How many grams of oxygen can be dissolved in 1.00 L of water at 20°C if the oxygen pressure is 2.00 atm? The Henry's law constant for oxygen at 20°C is 1.38×10^{-3} M/atm.

Consider a solution prepared by adding 50.0 g of sucrose ($C_{12}H_{22}O_{11}$) to 250.0 g of water to make 275.0 mL of solution.

8. If the vapor pressure of pure water at 25°C is 23.79 mm Hg, calculate the vapor pressure of the water over the solution at 25°C.

9. Calculate the freezing point of the solution.

10. Calculate the boiling point of the solution.

11. Calculate the osmotic pressure of the solution at 25°C.

ANSWERS
Exercises:
(E1) 1.10 M (E2) 0.0445 mol (E3) 33.9 g
(E4) 141 mL (E5) 0.113 mol NH_4^+; 0.0375 mol PO_4^{3-} (E6) 0.133 M
(E7) 2.7 g (E8) 0.13 mg (E9) 169 g/mol
(E10) Take 543 mL of the concentrated acid and add water to make 5.00 L of solution.
(E11) X_{solute} = 0.0450; 20.43%; 2.043 $\times 10^5$ ppm
(E12) 293 mm Hg; 107 mm Hg
(E13) 44.1°C (E14) 1.6 $\times 10^4$ g/mol

Self-Test
A. Multiple Choice:
1. c 2. b 3. b 4. c 5. c
6. a 7. c

B. Less than, Equal to, Greater than:
1. GT 2. GT 3. MI 4. MI 5. EQ
6. MI

C. Fill in the Blanks:
1. CH_3CH_2OH 2. NH_3 3. BY 4. M 5. tripled

D. True or False:
1. T 2. F 3. F 4. T 5. F
6. T

E. Problems:
1. 8.76% 2. 0.0177 3. 0.983 M 4. 0.999 m
5. 2.73 L 6. 6.97 M; 36.0%; 0.138 7. 0.0883 g 8. 23.54 mm Hg
9. −1.09°C 10. 100.30°C 11. 13.0 atm

11
Rate of Reaction

I. Meaning of Reaction Rate
  A. Definition of reaction rate:

_____

_____

  B. Mathematical expression for the reaction rate of the following reaction:

$$N_2O_5\,(g) \;\rightarrow\; 2\,NO_2\,(g) + \tfrac{1}{2}\,O_2\,(g)$$

_____

  C. Units
    1. The units are always concentration/time.
    2. The concentration is always in M or moles/liter (mol/L).
    3. The time can be in seconds, minutes, hours, days, or years.

II. Reaction Rate and Concentration
  A. Reaction rate decreases as the concentration of the reactant decreases.
  B. Rate expression and rate constant
    1. The rate expression for the reaction

$$N_2O_5\,(g) \;\rightarrow\; 2\,NO_2\,(g) + \tfrac{1}{2}\,O_2\,(g)$$

is

$$\text{rate} = k\,[N_2O_5]^m$$

    2. k
      a. k is called a rate constant.
      b. k is dependent on the nature of the reaction.
      c. k is dependent on temperature.
      d. k is independent of concentration.

C. Order of reaction involving a single reactant

1. It is the power to which the reactant is raised in the rate expression. In the rate expression described earlier, m is the order of the reaction.
2. It has to be determined experimentally.
3. It need *not* be the coefficient of the reactant in the balanced equation for the reaction.
4. Order versus concentration
   a. A zero order reaction is independent of concentration.
      Tripling the concentration multiplies the rate of the reaction by _____.
   b. In a first order reaction, the rate is directly proportional to the concentration.
      Tripling the concentration multiplies the rate of the reaction by _____.
   c. In a second order reaction, the rate is proportional to the square of the concentration of the reaction.
      Tripling the concentration multiplies the rate of the reaction by _____.

5. *Exercises*
   a. The initial rate of decomposition of a compound A to its products B and C,

$$A\,(g) \;\rightarrow\; B\,(g) + C\,(g)$$

was measured at a series of different concentrations with the following results:

[A] (M)	0.10	0.20	0.30	0.40
RATE (M/min)	0.20	0.56	1.04	1.60

Using these data, determine the order of the reaction

Solution: To determine the order of the reaction when you are given data on concentration and rates, choose two rates and their corresponding concentrations. We choose $[A]_1 = 0.10$, $rate_1 = 0.20$, $[A]_2 = 0.20$, and $rate_2 = 0.56$. Then we use the formula

$$\frac{rate_2}{rate_1} = \left(\frac{[A_2]}{[A_1]}\right)^m$$

where m is the order of the reaction.
Substituting, we get

$$\frac{0.56}{0.20} = \left(\frac{0.20}{0.10}\right)^m$$

$$2.8 = (2.0)^m$$

Unlike the Example 11.1 in the text that involved $2.2 = 1.5^m$, where m is easily determined to be 2, here m cannot be determined simply by inspection. We can only determine it by taking the log or ln of both sides.

$$\ln 2.8 = \ln 2.0^m$$

This is the same as

$$\ln 2.8 = m (\ln 2.0)$$

Solving for m we get

$$m = \frac{\ln 2.8}{\ln 2.0} = \frac{1.03}{0.69} = 1.5$$

The order of the reaction is 1.5.
Caution: ln 2.8/ln 2.0 is *not* the same as ln(2.8/2.0).

b. Calculate the rate constant for the preceding equation.
   Solution: Since we now have the order of the reaction, the rate expression for the reaction is

$$\text{rate} = k[A]^{1.5}$$

To determine k we can choose any rate and concentration from the data, making sure that they are from the same experiment. Let us choose 0.40 as [A]. The rate therefore is 1.60. Substituting, we get

$$1.60 = k (0.40)^{1.5}$$

$$k = \frac{1.60 \frac{\text{mol}}{L \cdot \text{min}}}{\left(0.40 \frac{\text{mol}}{L}\right)^{1.5}} = 6.3 \left(\frac{\text{mol}}{L}\right)^{-0.5} \text{min}^{-1}$$

c. What is the rate of the reaction if [A] is 0.50 M?
   Solution: Again, we use the equation

$$\text{rate} = k[A]^{1.5}$$

Substituting, we get

$$\text{rate} = \left[6.3 \left(\frac{\text{mol}}{L}\right)^{-0.5} \text{min}^{-1}\right] \left[0.50 \frac{\text{mol}}{L}\right]^{1.5} = 2.3 \frac{\text{mol}}{L \cdot \text{min}}$$

d. Consider the hypothetical decomposition

$$X (g) \rightarrow 2 Y (g) + 3 Z (g)$$

The initial rate of decomposition of X was measured at a series of different concentrations with the following results.

[X] (M)	0.10	0.20	0.30	0.40
RATE (M/s)	0.30	0.42	0.52	0.60

(1) What is the order of the reaction? **(E1)**

(2) Write the rate expression for the reaction. **(E2)**

(3) Calculate k for the reaction. **(E3)**

(4) What is the initial concentration if the rate is 0.90 mol/L·s? **(E4)**

D. Order of a reaction with two or more reactants
   1. The order of a reaction with respect to each reactant must be determined one at a time.
   2. The process is similar to the one discussed for one reactant. In choosing experiments from which to calculate for order, you have to choose those experiments where only the concentrations of the reactant in question change. All other concentrations must remain constant.
   3. Exercises
      a. Consider the hypothetical equation

$$A\,(g) \; + \; B\,(g) \quad \rightarrow \quad C\,(g) \; + \; D\,(g)$$

The following data were obtained for initial rates.

Expt.	[A] (M)	[B] (M)	Rate (mol/L·s)
1	0.200	0.200	2.5×10^{-3}
2	0.400	0.200	1.0×10^{-2}
3	0.400	0.400	1.0×10^{-2}

Calculate the order of the reaction with respect to A; with respect to B.
Solution: To calculate the order of the reaction with respect to A, we choose Experiments 1 and 2. In these experiments, the concentration of B is held constant. We use the formula

$$\frac{rate_2}{rate_1} = \left(\frac{[A_2]}{[A_1]}\right)^m$$

where m is the order of the reaction with respect to A. Substituting we get

$$\frac{1.0 \times 10^{-2}}{2.5 \times 10^{-3}} = \left(\frac{0.400}{0.200}\right)^m$$

$$4 = (2)^m$$

We see that m = 2. The reaction is second order with respect to A. We now choose experiments for reactant B. We choose Experiments 2 and 3 because in these experiments the concentration of A is held constant. Again, we substitute into the formula above.

$$\frac{1.0 \times 10^{-2}}{1.0 \times 10^{-2}} = \left(\frac{0.400}{0.200}\right)^n$$

$$1 = (2)^n$$

We see that n = 0, and hence for B, the reaction is zero order.

b. Write the rate law for the hypothetical reaction

$$2 A\,(g) + 3 B\,(g) \rightarrow C\,(g) + 2 D\,(g)$$

given the following initial rates. **(E5)**

Expt.	[A] (M)	[B] (M)	Rate (mol/L·s)
1	0.100	0.100	0.040
2	0.100	0.200	0.160
3	0.200	0.200	0.320

4. The overall order of a reaction
   a. When there are several reactants in a reaction, we refer to the order of the reaction "with respect to A" or B, and so on.
   b. The overall order is the sum of the orders with respect to each reactant.
   c. For the reaction

$$a\,A\,(g)\ +\ b\,B\,(g)\quad \rightarrow\quad products$$

   the rate expression is

   d. The overall order of the reaction is

III. Reactant Concentration and Time
   A. First order reactions
      1. Coordinates to obtain a straight-line plot: ln [X] (y-axis) versus time (x-axis)
      2. Equation
      We substitute into the general equation

$$y\ =\ mx\ +\ b$$

   where
      y   is the log of the concentration (ln [X])
      x   is time (t)

m (the slope) is the rate constant k. It is negative because the concentration decreases with time.

b is the y-intercept (what the concentration is when t is 0). It is called $\ln[X_0]$ (initial concentration).

The equation therefore is

$$\ln[X] = -kt + \ln[X_0]$$

Putting it in another form we get

$$\ln[X_0] - \ln[X] = kt$$

3. *Exercises:*

a. Paraldehyde ($C_6H_{12}O_3$) decomposes by a first order reaction at 225°C to acetaldehyde (C_2H_4O). The equation is

$$C_6H_{12}O_3\,(g) \rightarrow 3\,C_2H_4O\,(g)$$

The rate constant for this reaction at 225°C is $3.05 \times 10^{-4}\,s^{-1}$. If we start with a 0.250 M concentration of paraldehyde, what will its concentration be after 2.00 minutes?

Solution: Since the reaction is first order, we use the formula

$$\ln[X_0] - \ln[X] = kt$$

The initial concentration, $[X_0]$, is 0.250 mol/L. The rate constant, k, is $3.05 \times 10^{-4}\,s^{-1}$ and t is 2.00 min, which we must change to seconds so that the units will be consistent with those of k. Hence t = 1.20×10^2 s. The concentration, $[X]$, after 2.00 minutes is not known. We now plug into the equation above.

$$\ln 0.250 - \ln[X] = (3.05 \times 10^{-4}\,s^{-1})(1.20 \times 10^2\,s)$$

We take ln 0.250 for the left side and take the product of k and t for the right. We obtain

$$-1.386 - \ln[X] = 0.0366$$

Rearranging we get

$$\ln[X] = -1.386 - 0.0366 = -1.423$$

Now, we take the inverse ln of both sides. The inverse ln of $\ln[X]$ is $[X]$.

$$[X] = 0.241\,M$$

The concentration of paraldehyde after 2.00 minutes is 0.241 M.

b. The decomposition of sulfuryl chloride, SO_2Cl_2, into sulfur dioxide gas and chlorine gas is first order. The rate constant for the reaction is $2.2 \times 10^{-5}\,s^{-1}$ at 320°C.

   (1) Write the equation for the decomposition of sulfuryl chloride. **(E6)**

   (2) How long will it take to decompose 20.0% of the sulfuryl chloride? **(E7)**

   (3) If after heating a sample of sulfuryl chloride for six hours, 0.100 M is left, what was the initial concentration of the sulfuryl chloride? **(E8)**

4. Half-life

   a. Definition: _____

   _____

   b. Mathematical equation for determining the half-life of a first order reaction:

B. Zero order reaction

   1. Linear plot

     y-axis: _____

     x-axis: _____

2. Mathematical relationship of concentration with time:

3. Half-life equation:

C. Second order reaction
  1. Linear plot

  y-axis: _____

  x-axis: _____
  2. Mathematical relationship of concentration with time:

  3. Half-life equation:

IV. Activation Energy
  A. Characteristics
    1. It is a positive quantity.
    2. It depends on the nature of the reaction.
    3. It is independent of temperature and concentration.
    4. Its symbol is E_a.
  B. Activation energy versus ΔH

$$\Delta H = E_a - E_a'$$

where E_a' is the activation energy of the reverse reaction.

V. Reaction Rate versus Temperature
  A. Relation between k and T

$$\ln k_2 - \ln k_1 = \frac{E_a}{R}\left(\frac{1}{T_1} - \frac{1}{T_2}\right)$$

Remember: E_a has to be expressed in joules if 8.31 J/mol·K is the unit used for R. T has to be expressed in Kelvin. It is also a good idea to make the higher temperature T_2 to avoid negative logarithms.

B. *Exercises*

1. Nitrogen dioxide decomposes into nitrogen oxide and oxygen. At 319°C its rate constant is 0.498 L/mol·s, and at 383°C its rate constant is 4.74 L/mol·s. Calculate the activation energy for this reaction.

   Solution: The equation for the reaction is

$$2\,NO_2\,(g) \;\rightarrow\; 2\,NO\,(g)\,+\,O_2\,(g)$$

   The formula relating rate and temperature is the one given on the previous page. For T_2, we choose 383°C (656 K). The rate constant (k_2) associated with that temperature is 4.74 L/mol·s. T_1 is 319°C and k_1 is 0.498 L/mol·s. Subsituting, we get

$$\ln 4.74 \;-\; \ln 0.498 \;=\; \frac{E_a}{8.31}\left(\frac{1}{592} - \frac{1}{656}\right)$$

   Simplifying, we obtain

$$1.556 \;-\; (-0.697) \;=\; \frac{E_a}{8.31}\,(0.00169 \;-\; 0.00152)$$

   We do the arithmetic for both sides and get

$$2.253 \;=\; E_a\,(2.0 \times 10^{-5})$$

$$E_a \;=\; 1.1 \times 10^5 \;\frac{J}{mol}$$

2. Nitrogen oxide reacts with chlorine to form NOCl. Its activation energy is 17.5 kJ/mol. At 0°C, the rate constant for the reaction is 4.50 L^2/mol^2 · s. At what temperature will the rate be 8.00 L^2/mol^2 · s? **(E9)**

VI. Catalysis
    A. Definition of catalyst: _____

_____

    B. Types of catalysis
        1. Homogeneous

           Definition: _____

           _____

        2. Heterogeneous
           Definition: _____

           _____

    C. What are enzymes?

_____

VII. Reaction Mechanisms
    A. Rules to follow
        1. Rate expression for reactions in the mechanism
           The order of the reaction with respect to each reactant is its coefficient in the chemical equation for that step. Thus, if the equation

$$b\,B + a\,A \;\rightarrow\; c\,C$$

           were a reaction in a step of a given mechanism, the rate expression for it would be

$$\text{rate} = k\,[A]^{a}[B]^{b}$$

           Remember: This works only for the reactions that are part of a proposed mechanism.
        2. The slow step is the rate-determining step. If the following steps were proposed for a mechanism

$$b\,B + a\,A \;\rightarrow\; c\,C \qquad \text{(fast)}$$
$$x\,X + y\,Y \;\rightarrow\; z\,Z \qquad \text{(slow)}$$

           then the rate of the overall reaction would be

$$\text{rate} = k\,[X]^{x}[Y]^{y}$$

        3. The final rate expression must include only those species that appear in the balanced overall reaction. Unstable intermediates can usually be eliminated. Make use of the fact that reaction mechanisms many times have fast reactions in which the rate of the forward reaction equals the rate of the reverse reaction. If that is the case, the rate constant for the forward reaction can be denoted as k_1 while the rate constant for the reverse reaction is denoted as k_{-1}.
        4. Do not feel that you have to use all the steps in the mechanism to arrive at the rate equation for the overall reaction.

B. *Exercises*

1. The overall reaction

$$2\,NO\,(g)\ +\ O_2\,(g)\ \rightarrow\ 2\,NO_2\,(g)$$

is believed to take place in the following steps:

$$2\,NO\,(g)\ \rightleftharpoons\ N_2O_2\,(g) \qquad\qquad\text{(fast)}$$

$$N_2O_2\,(g)\ +\ O_2\,(g)\ \rightarrow\ 2\,NO_2\,(g) \qquad\text{(slow)}$$

Obtain a rate expression that corresponds to this proposed mechanism.
Solution: First, we determine the rate expression for the overall reaction. This will be the rate expression for the slow step. It is

$$\text{rate} = k_1\,[N_2O_2][O_2]$$

Both reactants have coefficients of 1, and thus their concentrations are also raised to the power 1. Now we see that our rate expression has the concentration of N_2O_2, which is not a species present in the overall equation. We try to eliminate the term $[N_2O_2]$ by using the rates of the fast reaction.

$$2\,NO\,(g)\ \rightleftharpoons\ N_2O_2\,(g)$$

We write the rate expression for both forward and reverse reactions, and equate them to each other since the rates are equal.

$$k_2\,[NO]^2 = k_{-2}\,[N_2O_2]$$

Thus,

$$[N_2O_2] = \frac{k_2\,[NO]^2}{k_{-2}}$$

Substituting into the rate expression that we wrote above, we get

$$\text{rate} = \frac{k_1 k_2}{k_{-2}}\,[NO]^2\,[O_2]$$

Combining all rate constants into one, and calling it k we have

$$\text{rate} = k\,[NO]^2\,[O_2]$$

2. The reaction
$$2\,NO\,(g)\ +\ Cl_2\,(g)\ \rightarrow\ 2\,NOCl\,(g)$$

is believed to take place in the following steps:

$$NO\,(g)\ +\ Cl_2\,(g)\ \rightleftharpoons\ NOCl_2\,(g) \qquad\qquad\text{(fast)}$$

$$NO \ (g) \ + \ NOCl_2 \ (g) \ \rightarrow \ 2 \ NOCl \ (g) \qquad \text{(slow)}$$

Obtain the rate expression corresponding to this mechanism. **(E10)**

SELF-TEST
A. Multiple Choice:
  1. Consider the reaction

$$2 \ N_2O_3 \ (g) \ + \ O_2 \ (g) \ \rightarrow \ 4 \ NO_2 \ (g)$$

Which of the following statements is correct?
  a. It is first order in N_2O_3.
  b. It is second order in N_2O_3.
  c. It is third order overall.
  d. It is both b and c.
  e. The order of the reaction cannot be predicted without more information.

  2. For the reaction
$$N_2 \ (g) \ + \ 3 \ H_2 \ (g) \ \rightarrow \ 2 \ NH_3 \ (g)$$

a catalyst is required. Addition of the catalyst
  a. lowers the activation energy.
  b. produces more H_2 (g) for the reaction.
  c. changes ΔH for the reaction.
  d. does none of the above.

  3. Which of the following steps appearing in a mechanism would be third order overall?
  a. $SO_2Cl_2 \ (g) \ \rightarrow \ SO_2 \ (g) \ + \ Cl_2 \ (g)$
  b. $2 \ NO \ (g) \ + \ O_2 \ (g) \ \rightarrow \ 2 \ NO_2 \ (g)$
  c. $2 \ N_2O \ (g) \ \rightarrow \ 2 \ N_2 \ (g) \ + \ O_2 \ (g)$
  d. $2 \ HI \ (g) \ \rightarrow \ H_2 \ (g) \ + \ I_2 \ (g)$

4. Which of the following statements is true for a zero order reaction?
   a. The rate constant equals zero.
   b. The activation energy equals zero.
   c. The rate is independent of concentration.
   d. The half-life is equal to 0.693/k.
   e. None of those.

5. Reaction rates increase with temperature. How many of the statements below explain the phenomenon?

   — Molecular velocities increase.
   — The average kinetic energy of the molecules increases.
   — The fraction of high energy molecules increases.
   — Collisions occur more frequently at high temperatures.

   a. 0   b. 1   c. 2   d. 3   e. 4

6. In a first order reaction with time measured in seconds (s) and concentration measured in mol/L, the rate constant k can have units of
   a. s   b. s^{-1} c. s·mol/L
   d. mol/L·s   e. none of these

7. A set of reactants reacts in two different ways to give a desired product (D) and an undesirable product (U). At 100°C, D and U are formed at the same rate. At 120°C, more D is formed than U. Which one of the following statements is true?
   a. The reaction forming U has the larger activation energy.
   b. The reaction forming D has the larger activation energy.
   c. Both reactions have the same activation energy.
   d. Not enough information is given.

8. To determine the rate constant of a suspected second order reaction, you would plot (where [X] is the concentration of the reactant and t is time)
   a. $[X^2]$ vs. t   b. [X] vs. t   c. ln [X] vs. t
   d. ln [X] vs. 1/t   e. 1/[X] vs. t

9. The value of the half-life of a first order reaction depends on
   (1) the initial concentration of the reactant.
   (2) the value of the rate constant.
   (3) the temperature of the reaction.
   (4) the ΔH of the reaction.
   (5) the final concentration of the reactant.

   a. (1),(2)   b. (2),(3)   c. (3),(4)   d. (4),(5)   e. (1),(3),(5)

10. Consider the following overall reaction.

$$NO_2 (g) + CO (g) \rightarrow NO (g) + CO_2 (g)$$

A proposed mechanism for this overall reaction is

$$2 NO_2 (g) \rightarrow NO (g) + NO_3 (g) \qquad \text{(slow)}$$

$$NO_3 (g) + CO (g) \rightleftharpoons NO_2 (g) + CO_2 (g) \qquad \text{(fast)}$$

If the proposed mechanism is valid, the rate expression for the overall reaction could be

a. $k [NO_3] [NO]$ **b.** $k [NO_2]^2$ **c.** $k [NO_2]$ **d.** $k [NO_2] [CO]$

B. True or False:

_____ **1.** In a zero order reaction, the reactant's concentration remains the same throughout the reaction.

_____ **2.** The unit for the rate of any reaction, regardless of order, is always concentration per unit time.

_____ **3.** The overall order of a reaction can be determined by adding the coefficients of the reactants in the balanced equation.

_____ **4.** The presence of a catalyst will not change the value of the rate constant, k, for the reaction, provided that the temperature is kept constant.

_____ **5.** For the reaction

$$A \rightarrow 2 B$$

the rate at which A is used up is half the rate at which B is produced.

_____ **6.** The order of a reaction with respect to a reactant can be 0.50.

_____ **7.** Consider two different reactions, 1 and 2, both done at the same temperature. If $E_{a_1} > E_{a_2}$, then $k_2 > k_1$.

_____ **8.** Any time a collision between two reactant molecules occurs, a reaction takes place.

_____ **9.** When two molecular compounds react, the bonds broken by the collision between the molecules are covalent bonds.

_____ **10.** The presence of a catalyst in a reaction changes both E_a and ΔH for that reaction.

_____ **11.** The presence of a catalyst speeds up a reaction.

_____ **12.** A catalyst does not affect the path a reaction takes.

_____ **13.** A catalyst shifts the equilibrium of a reaction.

_____ **14.** A catalyst lowers the activation energy of a given reaction.

_____ **15.** A catalyst gets consumed during a reaction.

C. Problems:

Consider the reaction

$$A + B_2 \rightarrow \text{products}$$

The following experimental data were obtained:

Expt.	[A] (M)	[B] (M)	Rate (mol/L·s)
1	0.100	0.100	8.0×10^{-3}
2	0.500	0.100	2.0×10^{-1}
3	0.100	0.500	4.0×10^{-2}

1. What is the order of the reaction with respect to each reactant?

2. What is the rate constant of the reaction?

3. The reaction is done at 22.0°C. It is determined that the activation energy of the reaction is 115 kJ/mol. What is the rate constant of the reaction at 27.0°C?

Consider the first order decomposition of A. The rate constant is 1.7×10^{-2} min^{-1}.
4. Calculate the rate of the reaction when the initial concentration of A is 0.200 M.

5. What percent of A will be used in one hour?

6. What is the half-life of the reaction?

ANSWERS
Exercises:

(E1) 1/2   (E2) rate = $k[X]^{\frac{1}{2}}$ (E3) $0.95\ mol^{\frac{1}{2}}/L^{\frac{1}{2}} \cdot s$

(E4) 0.90 M   (E5) rate = $k[A][B]^2$ (E6) $SO_2Cl_2\ (g) \rightarrow SO_2\ (g) + Cl_2\ (g)$

(E7) 2.8 hr   (E8) 0.16 M   (E9) 22°C

(E10) rate = $k[NO]^2[Cl_2]$

Self-Test
A. Multiple Choice:

1. e	**2.** a	**3.** b	**4.** c	**5.** e
6. b	**7.** a	**8.** e	**9.** b	**10.** b

B. True or False:

1. F	**2.** T	**3.** F	**4.** F	**5.** T
6. T	**7.** T	**8.** F	**9.** T	**10.** F
11. T	**12.** F	**13.** F	**14.** T	**15.** F

C. Problems:

1. second order in A, first order in B   2. $8.0\ L^2/mol^2 \cdot s$

3. $18\ L^2/mol^2 \cdot s$ 4. $3.4 \times 10^{-3}\ mol/L \cdot min$

5. 64 %   6. 41 min

12

Gaseous Chemical Equilibrium

I. The Equilibrium Constant Expression
A. General facts
1. For the general gas phase system

$$aA \text{ (g)} + bB \text{ (g)} \rightleftharpoons cC \text{ (g)} + dD \text{ (g)}$$

where A, B, C, and D are different reactants and products and a, b, c, and d are the coefficents used to balance the equation.

$$K = \frac{(P_C)^c (P_D)^d}{(P_A)^a (P_B)^b}$$

2. The equilibrium partial pressures of products (right side of the equation) appear in the numerator.
3. The equilibrium partial pressures of reactants (left side of the equation) appear in the denominator.
4. Each partial pressure is raised to a power equal to its coefficient in the balanced equation.

B. Changing the form of the chemical equation
1. The expression for K depends upon the form of the chemical equation written to describe the reaction at equilibrium.
2. When coefficients are changed

   a. Coefficient rule: _____

   _____

   _____

b. Example:
The following chemical equation is written to describe the reaction between hydrogen gas and chlorine gas to form hydrogen chloride gas.

$$H_2 \,(g) + Cl_2 \,(g) \; \rightleftharpoons \; 2\,HCl \,(g)$$

The equilibrium constant expression for the above equation is

$$K_1 \; = \; \frac{(P_{HCl})^2}{(P_{H_2})(P_{Cl_2})}$$

If the equation describing the same reaction is written

$$\tfrac{1}{2}\,H_2 \,(g) + \tfrac{1}{2}\,Cl_2 \,(g) \; \rightleftharpoons \; HCl \,(g)$$

then the equilibrium constant expression for this reaction is

$$K_2 \; = \; \frac{(P_{HCl})}{(P_{H_2})^{\frac{1}{2}}(P_{Cl_2})^{\frac{1}{2}}}$$

Relating K_1 to K_2 we get
$$K_2 \; = \; \sqrt{K_1}$$

3. When the direction of the equation is reversed

    a. Reciprocal rule: _____

_____

    b. Example:
Consider again the reaction at equilibrium between hydrogen gas and chlorine gas to form hydrogen chloride gas. If the equation is written

$$2\,HCl \,(g) \; \rightleftharpoons \; H_2 \,(g) + Cl_2 \,(g)$$

then the equilibrium expression is

$$K_3 \; = \; \frac{(P_{H_2})(P_{Cl_2})}{(P_{HCl})^2}$$

Relating K_3 to K_1, written earlier, we see that

$$K_3 \; = \; \frac{1}{K_1}$$

C. Adding chemical equations
    1. Rule of multiple equilibria: _____

_____

2. Mathematical relationship
If

$$\text{Reaction 3} = \text{Reaction 1} + \text{Reaction 2}$$

then

$$K\,(\text{reaction 3}) = K\,(\text{reaction 1}) \times K\,(\text{reaction 2})$$

D. Heterogeneous Equilibria
   1. Homogeneous systems are those in which all reactants and products are in one phase. We have been looking at equilibrium expressions for these systems — all in the gas phase.
   2. Heterogeneous systems are those in which more than one phase is present.
   3. Equilibria in heterogeneous systems
      The position of the equilibrium is independent of the amount of solid or pure liquid, as long as some is present.
   4. Writing equilibrium expressions for heterogeneous systems:
      a. Terms for solids or pure liquids need not appear in the expression for K.
      b. Ions or molecules in solution appear in the expression for K. Their concentration in molarity [X] is used.

E. *Exercises*
   1. Write the equilibrium constant expression K for the reaction between sulfur dioxide gas and oxygen to produce sulfur trioxide gas when it reaches equilibrium.
   Solution: One equation for the reaction is

$$2\,SO_2\,(g) + O_2\,(g) \rightleftharpoons 2\,SO_3\,(g)$$

As you can see, it is necessary to be able to write out equations for reactions and *balance* them before you can write out the expression for K. The expression for K is

$$K = \frac{(P_{SO_3})^2}{(P_{SO_2})^2\,(P_{O_2})}$$

If you balanced the same equation in the following way

$$SO_2\,(g) + \tfrac{1}{2}\,O_2\,(g) \rightleftharpoons SO_3\,(g)$$

then your expression for K would be

$$K = \frac{(P_{SO_3})}{(P_{SO_2})\,(P_{O_2})^{\frac{1}{2}}}$$

Both expressions for K are correct, provided you include the corresponding chemical equation.
   2. Write the equilibrium expression for K for the following reactions:
      a. $4\,NH_3\,(g) + 5\,O_2\,(g) \rightleftharpoons 4\,NO\,(g) + 6\,H_2O\,(g)$ **(E1)**

b. $Pb(NO_3)_2$ (s) $\rightleftharpoons$ PbO (s) + 2 NO_2 (g) + $\frac{1}{2} O_2$ (g)   **(E2)**

II. Determination of K
There are three ways to calculate K.
A. Given the partial pressures of products and reactants at equilibrium
   1. Pressures do not have to be expressed in a specific pressure unit as long as all the units are the same.
   2. *Exercises*
     a. An equilibrium mixture for the system

$$H_2 \ (g) + CO_2 \ (g) \ \rightleftharpoons \ H_2O \ (g) + CO \ (g)$$

at 990°C is analyzed. It is found that the partial pressures of H_2, CO_2, H_2O, and CO are 464 mm Hg, 1.6 atm, 1.1 atm, and 1.4 atm, respectively. What is K for the reaction at 990°C?

Solution: First, we change 464 mm Hg to atm, so all the partial pressures are in atmospheres.

$$464 \ mm \ Hg \ \times \ \frac{1 \ atm}{760 \ mm \ Hg} = 0.611 \ atm$$

Then we write the equilibrium expression for K.

$$K = \frac{(P_{H_2O}) \, (P_{CO})}{(P_{H_2}) \, (P_{CO_2})}$$

Finally, we simply plug in the given partial pressures.

$$K = \frac{(1.1) \, (1.4)}{(0.611) \, (1.6)} = 1.6$$

    b. At a certain temperature, the equilibrium mixture of PCl_3, PCl_5, and Cl_2, has the following partial pressures: $P_{Cl_2} = 0.11$ atm, $P_{PCl_3} = 0.17$ atm, $P_{PCl_5} = 0.37$ atm. Calculate K for the reaction

$$PCl_3 \ (g) + Cl_2 \ (g) \ \rightleftharpoons \ PCl_5 \ (g)   \textbf{(E3)}$$

c. Solid ammonium chloride decomposes to ammonia gas and hydrogen chloride gas. At 78°C, the equilibrium mixture contained in a 5.00-L flask is analyzed and found to have 1.12 g of ammonium chloride, 4.13 g of ammonia gas, and 2.46 g of hydrogen chloride gas.

   (1) Write the equation for the decomposition of one mole of ammonium chloride.

   (2) Write the equilibrium expression for K.

   (3) Calculate K at 78°C. **(E4)**

B. Given the initial partial pressure of one species and the equilibrium partial pressure of another
   1. In order to solve problems like these, you must do the following:
      a. Carefully distinguish between the initial or original partial pressure (P_o), and the partial pressure at equilibrium (P_{eq}).
      b. Write a balanced equation for the reaction. A balanced equation relates the number of moles as well as the partial pressures of the products and the reactants to each other by the coefficients.
      c. Construct a table to manage your data systematically.
   2. *Exercises*
      a. Before equilibrium is reached at a certain temperature, a one-liter flask contains only SO_3 gas at a pressure of 0.850 atm. What is K for the reaction

$$2\,SO_3\,(g) \;\rightleftharpoons\; O_2\,(g) + 2\,SO_2\,(g)$$

if, at equilibrium, oxygen with a partial pressure of 0.226 atm is present?
Solution: In solving this type of problem (and in fact in solving any type of equilibrium problem), you should get into the habit of constructing a table. We construct a blank one using the SO_3 decomposition as an example.

	$2\,SO_3$	$\rightleftharpoons$	O_2	$+$	$2\,SO_2$
P_o					
ΔP					
P_{eq}					

We now put the given information into the table:

	$2\,SO_3$	$\rightleftharpoons$	O_2	$+$	$2\,SO_2$
P_o (atm)	0.850		0		0
ΔP (atm)					
P_{eq} (atm)			0.226		

To fill in the row for ΔP, we start by putting a "+" for species that are products (because we are making them) and a "−" for species that are reactants (because we are consuming them). The table now looks like this.

	$2\,SO_3$	$\rightleftharpoons$	O_2	$+$	$2\,SO_2$
P_o (atm)	0.850		0		0
ΔP (atm)	−		+		+
P_{eq} (atm)			0.226		

We continue filling in the blanks in the ΔP row. We have enough data to determine ΔP for oxygen, which is $0.226 = 0 + \Delta P$ or 0.226 atm. Our table now looks like this:

	$2\,SO_3$	$\rightleftharpoons$	O_2	$+$	$2\,SO_2$
P_o (atm)	0.850		0		0
ΔP (atm)	−		+ 0.226		+
P_{eq} (atm)			0.226		

Recall too that pressure is directly proportional to the number of moles of a gas. The partial pressure of SO_2 is therefore twice that of oxygen at equilibrium, since SO_2 has twice as many moles as oxygen in the reaction. Thus if ΔP for oxygen is 0.226 atm, the change in partial pressure for SO_2 must be twice that or 0.452 atm. The same reasoning holds true for ΔP of SO_3. The table now looks like this:

	$2\,SO_3$	$\rightleftharpoons$	O_2	$+$	$2\,SO_2$
P_o (atm)	0.850		0		0
ΔP (atm)	− 0.452		+ 0.226		+ 0.452
P_{eq} (atm)			0.226		

It now becomes a matter of arithmetic to fill in P_{eq} for SO_3

$$0.850 - 0.452 = 0.398$$

and P_{eq} for SO_2

$$0 + 0.452 = 0.452$$

The completed table now is

	2 SO₃	⇌	O₂	+	2 SO₂
P_o (atm)	0.850		0		0
ΔP (atm)	− 0.452		+ 0.226		+ 0.452
P_{eq} (atm)	0.398		0.226		0.452

Calculating K has now boiled down to a simple plug-in to the equilibrium expression using the data in the P_{eq} row.

$$K = \frac{(P_{O_2})(P_{SO_2})^2}{(P_{SO_3})^2} = \frac{(0.226)(0.452)^2}{(0.398)^2} = 0.291$$

b. Suppose that at another temperature we start with pure SO_3 at a pressure of 0.500 atm. At equilibrium, SO_2 has a partial pressure of 0.350 atm. What is K at that temperature? **(E5)**

C. Given the initial partial pressure of one species and how much of that species was consumed

1. Recall again that pressure is proportional to the number of moles (n) of a gas. If 25% of the gas is said to be consumed, then the pressure of the gas decreases by 25% of its original value. Read carefully to see whether the problem says that 25% was used up or 25% is left. That changes things a lot!

2. *Exercises*

   a. Consider the reaction

$$H_2 \text{ (g)} + I_2 \text{ (g)} \rightleftharpoons 2\,HI \text{ (g)}$$

   Suppose that we start with hydrogen gas at 0.2000 atm pressure and iodine gas at 0.1000 atm pressure. When equilibrium is reached at a certain temperature, 48.0% of the hydrogen gas will have been consumed. Calculate K at that temperature.

Solution: Again, we start with our table:

	H_2	+	I_2	$\rightleftharpoons$	2 HI
P_0 (atm)	0.2000		0.1000		0
ΔP (atm)					
P_{eq} (atm)					

Since 48.0% of the hydrogen was used, ΔP will be 48.0% of 0.2000 or 0.0960 atm. The change, ΔP for I_2, will be the same since for every mole of hydrogen used up, an equal amount of iodine is also used up. We know this relationship by looking at the equation and noting that one mole of hydrogen reacts with one mole of iodine. (You do *not* take 48.0% of 0.1000 atm.) The change, ΔP for HI, can be obtained by noting that in the equation, two moles of HI are produced for every mole of hydrogen, or 2 atm of HI are equivalent to one atm of hydrogen. Thus

$$0.0960 \text{ atm } H_2 \times \frac{2 \text{ atm HI}}{1 \text{ atm } H_2} = 0.192 \text{ atm}$$

After filling in the ΔP row with the data that we just obtained, and then doing the arithmetic within each column to fill in P_{eq}, our table looks like this:

	H_2	+	I_2	$\rightleftharpoons$	2 HI
P_0 (atm)	0.2000		0.1000		0
ΔP (atm)	− 0.0960		− 0.0960		+ 0.192
P_{eq} (atm)	0.1040		0.0040		0.192

Thus K can be calculated.

$$K = \frac{(P_{HI})^2}{(P_{I_2})(P_{H_2})} = \frac{(0.192)^2}{(0.0040)(0.1040)} = 89$$

b. Suppose that at another temperature we start with hydrogen gas at a partial pressure of 0.450 atm and iodine gas at a partial pressure of 0.350 atm. When equilibrium is reached, we find that 30.0% of the iodine has reacted. Calculate K for the reaction

$$H_2 \text{ (g)} + I_2 \text{ (g)} \rightleftharpoons 2\,HI \text{ (g)}$$

at that temperature. **(E6)**

III. Applications of the Equilibrium Constant
    A. Feasibility of the reaction
        1. Qualitatively, a very small K means that a reaction is not likely to happen.
        2. Qualitatively, a very large K means that a reaction is feasible.
    B. The direction of the reaction
        1. Reaction quotient – Q
        The expression for Q is the same as that for K, except that in Q original partial pressures (P_o) instead of equilibrium partial pressures (P_{eq}) are used.
          a. $Q > K$
            (1) This means that the denominator must become larger to reach equilibrium; hence, more reactants (species on the left-hand side) are produced.
            (2) The reaction proceeds from right to left ($\leftarrow$).
          b. $Q < K$
            (1) This means that the denominator must become smaller to reach equilibrium; hence, more products (species on the right-hand side) are produced.
            (2) The reaction proceeds from left to right ($\rightarrow$)
          c. $Q = K$
          The reaction is at equilibrium.
        2. Exercises
          a. For the reaction

$$N_2 \text{ (g)} + O_2 \text{ (g)} \rightleftharpoons 2\,NO \text{ (g)}$$

K is 6.2×10^{-4} at 2000°C. In which direction will the reaction proceed at 2000°C if original partial pressures are: $P_{N_2} = 0.52$ atm, $P_{O_2} = 0.124$ atm, and $P_{NO} = 0.020$ atm?
Solution: First we write the expression for Q.

$$Q = \frac{(P_{NO})^2}{(P_{N_2})(P_{O_2})}$$

Plugging in the given partial pressures, we get

$$Q = \frac{(0.020)^2}{(0.52)(0.124)} = 6.2 \times 10^{-3}$$

Since K is 6.2×10^{-4}, we see that $Q > K$ and the reaction goes from right to left ($\leftarrow$).

b. At the same temperature, in which direction will the reaction proceed if, in a 2.00-L flask, the partial pressures of NO, N_2, and O_2 are 2.4 mm Hg, 0.20 atm, and 0.60 atm respectively? **(E7)**

C. Equilibrium partial pressures

 1. Given K and equilibrium partial pressures of all species but one
    a. In such a case, the problem is a simple plug-in.
    b. *Exercises*
       (1) Consider the reaction

$$N_2O \text{ (g)} + \tfrac{1}{2}O_2 \text{ (g)} \rightleftharpoons 2\,NO \text{ (g)}$$

If $K = 8.3 \times 10^{-13}$ at 25°C, what is P_{NO} if P_{N_2O} is 0.0035 atm and P_{O_2} is 0.0027 atm?

Solution: First we write the equilibrium expression for the reaction.

$$K = \frac{(P_{NO})^2}{(P_{N_2O})(P_{O_2})^{\frac{1}{2}}}$$

Next we simply plug in all our information.

$$8.3 \times 10^{-13} = \frac{(P_{NO})^2}{(0.0035)(0.0027)^{\frac{1}{2}}}$$

Solving for P_{NO}, we get

$$(P_{NO})^2 = 1.5 \times 10^{-16}$$
$$P_{NO} = 1.2 \times 10^{-8} \text{ atm}$$

(2) For the reaction

$$N_2 (g) + 3 H_2 (g) \rightleftharpoons 2 NH_3 (g)$$

K is 8.7×10^{-8} at 25°C. How many milligrams of ammonia are in a ten-liter flask if at equilibrium $P_{N_2} = 2.00$ atm and $P_{H_2} = 0.80$ atm? **(E8)**

2. Given only the original pressures of all the species
Exercises
a. Consider the reaction

$$H_2 (g) + I_2 (g) \rightleftharpoons 2 HI (g)$$

The equilibrium constant for this reaction at 445°C is 50.2. If we start with the following partial pressures: $P_{H_2} = 0.100$ atm, $P_{I_2} = 0.100$ atm, and $P_{HI} = 0.050$ atm, what are the partial pressures at equilibrium for all three species? *Solution:* Again, we start by making a table.

	H_2	+	I_2	$\rightleftharpoons$	2 HI
P_o (atm)	0.100		0.100		0.050
ΔP (atm)	−		−		+
P_{eq} (atm)					

To fill in the rest of the row for ΔP, we start by putting an x for H_2. Since one mole of H_2 reacts with one mole of I_2 giving 2 moles of HI, ΔP for I_2 will also be x, while ΔP for HI will be 2x. The table now looks like this:

	H_2	+	I_2	$\rightleftharpoons$	2 HI
P_o (atm)	0.100		0.100		0.050
ΔP (atm)	− x		− x		+ 2x
P_{eq} (atm)					

To fill in the row for P_{eq}, we do the arithmetic indicated for each column. Thus,

	H_2	$+$	I_2	$\rightleftharpoons$	2 HI
P_0 (atm)	0.100		0.100		0.050
ΔP (atm)	$-x$		$-x$		$+2x$
P_{eq} (atm)	$0.100 - x$		$0.100 - x$		$0.050 + 2x$

Now we plug in P_{eq} for each species into the equilibrium expression

$$K = \frac{(P_{HI})^2}{(P_{I_2})(P_{H_2})}$$

and get

$$50.2 = \frac{(0.050 + 2x)^2}{(0.100 - x)(0.100 - x)}$$

$$= \frac{(0.050 + 2x)^2}{(0.100 - x)^2}$$

We could continue to solve for x by using the quadratic equation. In this particular case, though, it is simpler to take the square root of both sides. That gives

$$7.09 = \frac{(0.050 + 2x)}{(0.100 - x)}$$

$$0.709 - 7.09x = 0.050 + 2x$$

$$x = 0.0725 \, \text{atm}$$

This is *not* the answer. It is merely the *change*. Substitute 0.0725 atm for x in the entries in the P_{eq} row. Thus

$$P_{H_2} = 0.100 - 0.0725 = 0.028 \, \text{atm}$$

$$P_{I_2} = 0.100 - 0.0725 = 0.028 \, \text{atm}$$

$$P_{HI} = 0.050 + 2(0.0725) = 0.195 \, \text{atm}$$

Note: Do not be alarmed if you get a negative answer for x, as long as your equilibrium partial pressures end up being positive. However, if the equilibrium partial pressures are negative, either you have chosen the wrong root in your quadratic equation or something in your algebra is incorrect.

b. For the reaction

$$H_2 \, (g) + CO_2 \, (g) \rightleftharpoons H_2O \, (g) + CO \, (g)$$

K is 1.6 at 990°C. Calculate the partial pressures at equilibrium of all the species if we start with 0.250 atm of H_2, 0.250 atm of CO_2, and 0.100 atm of H_2O. **(E9)**

IV. Effect of Changes in the Conditions upon an Equilibrium System

A. LeChatelier's principle
Statement of the principle:

_____

_____

B. Qualitative Changes
  1. Adding or subtracting a product or reactant
    a. If the species added or subtracted is a liquid or a solid, there is no effect on equilibrium.
    b. If the species added or subtracted is a gas, then:
      (1) Adding a species will shift the direction away from the species.
      (2) Subtracting a species will shift the direction toward the species.
  2. Change in volume
    a. When the volume is decreased, the reaction takes place in the direction that decreases the total number of moles of gas.
    b. When the volume is increased, the reaction takes place in the direction that increases the total number of moles of gas.
    c. If in the balanced equation there is no increase or decrease in the total number of moles of gas, a volume change will not affect the equilibrium.
  3. Change in pressure
    a. An increase in pressure (by compression) shifts the equilibrium in the direction of decrease in the total number of moles of gas.
    b. A decrease in pressure (by expansion) shifts the equilibrium in the direction of increase in the total number of moles of gas.
  4. Change in temperature
    a. If the forward reaction (left to right) is endothermic, then:
      (1) An increase in temperature causes K to become larger.
      (2) A decrease in temperature causes K to become smaller.

b. If the forward reaction (left to right) is exothermic, then:

(1) An increase in temperature causes K to become smaller.

(2) A decrease in temperature causes K to become larger.

5. *Exercises*

   a. Consider the reaction

$$2\,SO_2\,(g) + O_2\,(g) \;\rightleftharpoons\; 2\,SO_3\,(g)$$

   In which direction will equilibrium shift if:

   (1) more oxygen is added?

   Solution: We know that adding a species shifts the equilibrium away from the species. Thus, the shift of the equilibrium is from left to right, that is, $\rightarrow$.

   (2) SO_3 is removed?

   Solution: We know that removing a species shifts the equilibrium toward the species being removed. Then the shift is from left to right, that is, $\rightarrow$.

   (3) the volume is increased?

   Solution: The shift is toward the side that has more moles of gas. Hence, the shift is to the left, that is, $\leftarrow$.

   (4) the pressure is increased by decreasing the volume?

   Solution: The shift is toward the side with fewer moles of gas. Hence, the shift is to the right, that is, $\rightarrow$.

   (5) the temperature is increased?

   Solution: First, we have to find ΔH for the reaction. Solving for ΔH using Table 8.3 in Chapter 8 of the text, we find that $\Delta H = -197.8$ kJ. The reaction is exothermic and an increase in temperature tends to make K smaller; hence more reactants are produced (larger denominator), and the shift is to the left, that is, $\leftarrow$.

b. Consider the reaction

$$COCl_2\,(g) \;\rightleftharpoons\; CO\,(g) + Cl_2\,(g) \qquad \Delta H > 0$$

   In which direction will equilibrium shift if:

   (1) the amount of CO is decreased?

   (2) the amount of Cl_2 is increased?

   (3) the volume is decreased?

(4) the pressure is decreased by increasing the volume?

(5) the temperature is increased? **(E10)**

C. Quantitative Changes
  1. Changes in Pressure
    a. Consider again the equation

$$H_2 \text{ (g)} + I_2 \text{ (g)} \rightleftharpoons 2\,HI \text{ (g)}$$

where K is 45.2 at 445°C. At equilibrium the partial pressures of the species are: $P_{H_2} = 0.204$ atm, $P_{I_2} = 0.00400$ atm, and $P_{HI} = 0.192$ atm. What will the partial pressures be when equilibrium is reestablished, if the pressure of iodine is increased to be the same as that of hydrogen?

Solution: When the equilibrium of the system is disturbed, the equilibrium partial pressures before the stress was imposed on the system become the original partial pressures. If we put this data in a table, it will look like this:

	H_2	+	I_2	$\rightleftharpoons$	2 HI
P_o (atm)	0.204		0.204		0.192
ΔP (atm)	$-x$		$-x$		$+2x$
P_{eq} (atm)	$0.204 - x$		$0.204 - x$		$0.192 + 2x$

Substituting into the equilibrium expression

$$K = \frac{(P_{HI})^2}{(P_{I_2})(P_{H_2})}$$

we get

$$45.2 = \frac{(0.192 + 2x)^2}{(0.204 - x)(0.204 - x)} = \frac{(0.192 + 2x)^2}{(0.204 - x)^2}$$

Taking the square root of both sides gives

$$6.72 = \frac{(0.192 + 2x)}{(0.204 - x)}$$

$$1.37 - 6.72\,x = 0.192 + 2x$$

$$x = 0.135 \text{ atm}$$

The equilibrium concentrations therefore are:

$$P_{HI} = 0.192 + 2(0.135) = 0.462 \text{ atm}$$

$$P_{H_2} = P_{I_2} = 0.204 - 0.135 = 0.069 \text{ atm}$$

b. Consider the reaction

$$CO \text{ (g)} + H_2O \text{ (g)} \rightleftharpoons CO_2 \text{ (g)} + H_2 \text{ (g)}$$

K for this reaction at 827°C is 1.00. At 827°C, the partial pressures of the different species are: P_{CO_2} = 0.200 atm, P_{H_2} = 0.600 atm, P_{CO} = 0.300 atm, and P_{H_2O} = 0.400 atm. If the partial pressure of the water is increased to 0.500 atm, what are the partial pressures of all the species when equilibrium is reestablished? **(E11)**

2. Changes in Temperature
    a. A change in temperature not only shifts the direction of equilibrium, but it also changes the value of K.
    b. The van Hoff't equation relates equilibrium constants to temperature. The equation is

$$\ln K_2 - \ln K_1 = \frac{\Delta H°}{R} \left(\frac{1}{T_1} - \frac{1}{T_2} \right)$$

Remember that $\Delta H°$ is the standard enthalpy change for the forward reaction as written. It has to be expressed in joules if R is expressed as 8.31 J/mol-K.
    c. *Example:*
    Consider again the reaction between hydrogen and iodine gases:

$$H_2 \text{ (g)} + I_2 \text{ (g)} \rightleftharpoons 2 HI \text{ (g)}$$

where K is 45.2 at 445°C. Calculate K at room temperature (23°C).

Solution: We will use the van Hoff't equation and note that we need $\Delta H°$ for the forward reaction as written, which is

$$H_2\,(g)\ +\ I_2\,(g)\ \rightarrow\ 2\,HI\,(g)$$

Using Table 8.3 of your text, we find that $\Delta H°$ for the reaction is +53.0 kJ. Substituting our known values for K_2 and T_2, we get

$$\ln 45.2\ -\ \ln K_1\ =\ \frac{53.0 \times 10^3}{8.31}\left(\frac{1}{296}\ -\ \frac{1}{718}\right)$$

Working the right side and determining the value for ln 45.2 for the left side, we obtain

$$3.81\ -\ \ln K_1\ =\ 12.66$$

Solving for K_1

$$\ln K_1\ =\ 3.81\ -\ 12.66$$
$$K_1\ =\ 1.43 \times 10^{-4}$$

Note how much smaller the equilibrium constant of an endothermic reaction becomes when temperature is decreased.

d. *Exercise:*
For the reaction

$$CO_2\,(g)\ +\ H_2\,(g)\ \rightleftharpoons\ CO\,(g)\ +\ H_2O\,(g)$$

K at 827°C is 1.00. At what temperature will K be 0.500? **(E12)**

SELF-TEST
A. Multiple Choice:

1. For the following system at equilibrium

$$C \text{ (s)} + H_2O \text{ (g)} \rightleftharpoons CO \text{ (g)} + H_2 \text{ (g)}$$

increasing the volume of the container
a. forms more C (s).
b. forms more CO (g) and H_2 (g).
c. has no effect on equilibrium.

2. For the reaction

$$A \text{ (g)} + B \text{ (g)} \rightarrow 2C \text{ (g)}$$

K is 9.0 at 35°C. The original partial pressures of the species are: $P_A = 0.2$ atm, $P_B = 0.4$ atm, and $P_C = 0.15$ atm. Which of the following statements is true?
a. The reaction will move to the left to achieve equilibrium.
b. The reaction will move to the right to achieve equilibrium.
c. The mixture of gases is at equilibrium.
d. The mixture cannot achieve equilibrium unless more gas is added.

3. The exothermic reaction

$$X \text{ (g)} + Y \text{ (g)} \rightarrow 2Z \text{ (g)}$$

reaches equilibrium at 200°C. How many of the following statements are true?
— If the mixture is transferred to a reaction flask of twice the volume, the amount of products and reactants will not change.
— Addition of an appropriate catalyst will result in the formation of a greater amount of Z (g).
— Lowering the temperature to 100°C will increase the amount of Z (g).
— Addition of helium will have little or no effect on the amount of Z (g) obtained.
a. 0 **b.** 1 **c.** 2 **d.** 3 **e.** 4

4. How is the following equilibrium shifted by cooling?

$$CuS \text{ (s)} + H_2 \text{ (g)} \rightleftharpoons H_2S \text{ (g)} + Cu \text{ (s)} \qquad \Delta H = 28 \text{ kJ}$$

a. Hydrogen sulfide gas is formed.
b. Hydrogen sulfide gas is consumed.
c. There is no change.

5. Three chemical systems are each in equilibrium.
(1) $2CO \text{ (g)} + O_2 \text{ (g)} \rightleftharpoons 2CO_2 \text{ (g)}$
(2) $H_2 \text{ (g)} + Cl_2 \text{ (g)} \rightleftharpoons 2HCl \text{ (g)}$
(3) $2HI \text{ (g)} \rightleftharpoons H_2 \text{ (g)} + I_2 \text{ (s)}$

The pressure on each of these systems is increased so that the volume decreases from 20 L to 10 L. The equilibrium shift is correctly represented as

a. (1) $\rightarrow$ (2) $\rightarrow$ (3)   no change
b. (1)   no change   (2) $\leftarrow$ (3) $\leftarrow$
c. (1) $\rightarrow$ (2)   no change   (3) $\rightarrow$
d. (1) $\leftarrow$ (2) $\rightarrow$ (3)   no change
e. (1) $\rightarrow$ (2)   no change   (3)   no change

6. The gaseous reaction

$$N_2 \, (g) \; \rightleftharpoons \; 2 \, N \, (g) \qquad \Delta H = +941 \, kJ$$

is in equilibrium at 1200 K. Which of the following operations will shift the reaction to the left?

(1) Addition of nitrogen (N_2) gas
(2) Increasing the temperature
(3) Increasing the pressure

a. (1)   b. (2),(3)   c. (1),(3)   d. (3)   e. none of those

7. For the reaction

$$A \, (g) + 2 \, B \, (g) \; \rightarrow \; 3 \, C \, (g)$$

K is 1.0 at 500°C. The original partial pressures of the species at 500°C are: $P_A = P_B = P_C = 0.2$ atm. Which of the following statements is true?

a. The partial pressure of A will increase as equilibrium is reached.
b. The partial pressure of B will increase as equilibrium is reached.
c. The partial pressure of C will increase as equilibrium is reached.
d. The system is already at equilibrium.
e. None of the above statements represents reality.

8. At 25°C

$$2 \, NO \, (g) \; \rightleftharpoons \; N_2 \, (g) + O_2 \, (g) \qquad K = 1 \times 10^{30}$$
$$2 \, H_2O \, (g) \; \rightleftharpoons \; 2 \, H_2 \, (g) + O_2 \, (g) \qquad K = 5 \times 10^{-82}$$
$$2 \, CO \, (g) + O_2 \, (g) \; \rightleftharpoons \; 2 \, CO_2 \, (g) \qquad K = 2.5 \times 10^{91}$$

The compound most likely to dissociate and give O_2 (g) at 25°C is
a. NO   b. H_2O c. CO_2 d. CO

9. What is the equilibrium expression for the following reaction?

$$2 \, SO_2 \, (g) + O_2 \, (g) \; \rightleftharpoons \; 2 \, SO_3 \, (g)$$

a. $\dfrac{(P_{SO_3})^2}{(P_{SO_2})^2 \, (P_{O_2})}$ b. $\dfrac{(P_{SO_3})^2}{(P_{SO_2})^2 + (P_{O_2})}$ c. $\dfrac{2 \, (P_{SO_3})}{2 \, (P_{SO_2}) \, (P_{O_2})}$

d. $\dfrac{(P_{SO_2})^2 \, (P_{O_2})}{(P_{SO_3})^2}$ e. $\dfrac{4 \, (P_{SO_3})^2}{(P_{SO_2})^2 \, (P_{O_2})}$

10. For the reaction

$$CO\ (g) + H_2O\ (g) \rightleftharpoons CO_2\ (g) + H_2\ (g)$$

K is 1.00 at about 1100 K. Which one of the following statements must always be true concerning this reaction at equilibrium?

a. $P_{CO} = P_{H_2O} = P_{CO_2} = P_{H_2}$

b. $P_{CO} \times P_{H_2O} = P_{CO_2} \times P_{H_2}$

c. $P_{CO} = P_{H_2O}$ and $P_{CO_2} = P_{H_2}$

d. $P_{CO} \times P_{H_2O} = P_{CO_2} \times P_{H_2} = 1.00$

e. All of the above are true.

B. Circle the correct answer:

Consider the system

$$N_2\ (g) + 3\,H_2\ (g) \rightleftharpoons 2\,NH_3\ (g) \qquad \Delta H = -92.2\ kJ$$

Show the effect on the partial pressure of H_2 and on the numerical value of K by circling the letters D for a decrease, I for an increase and NC for no change.

	P_{H_2}			K		
1. adding NH₃	D	I	NC	D	I	NC
2. doubling the volume	D	I	NC	D	I	NC
3. increasing T	D	I	NC	D	I	NC
4. adding Ar (g)	D	I	NC	D	I	NC
5. adding N₂ (g)	D	I	NC	D	I	NC

C. Fill in the blanks:

The following questions are about the following hypothetical reaction:

$$2\,R\ (g) + S\ (\ell) \rightleftharpoons T\ (g) + 2\,U\ (g) \qquad K = 1$$

_____ **1.** What is the partial pressure of R at equilibrium if the equilibrium partial pressure of T is 1 atm and that of U is 2 atm.

_____ **2.** In which direction will the reaction proceed ($\rightarrow$, $\leftarrow$, or no change) if the volume of S is doubled?

_____ **3.** It is determined that K increases when the temperature is increased. What is the sign of $\Delta H°$ for the reaction?

_____ 4. In which direction will the reaction proceed if Q = 2?
(→, ←, can't tell)

_____ 5. What is K (equilibrium constant) at the same temperature if the
pressure of R is doubled from 1 atm to 2 atm?

D. Problems:

Consider the hypothetical reaction

$$A \text{ (g)} + B \text{ (g)} \rightleftharpoons 2C \text{ (g)} \qquad \Delta H < 0$$

1. Write the equilibrium expression for the reaction.

2. Calculate K at a certain temperature if P_A = 0.250 atm, P_B = 0.135 atm, and P_C = 0.500 atm.

3. For another temperature, calculate K if we start with a 5.00-liter flask in which P_A = 0.300 atm and P_B = 0.750 atm. At equilibrium, 25.0% of A has been used up.

4. If K for the reaction at 100°C is 50.0, and we start with a 1.00-L flask where:

$P_A = P_B = 0.250$ atm, and $P_C = 0.500$ atm

a. Determine the direction of the reaction.

b. Calculate the partial pressures at equilibrium of A, B, and C at 100°C.

c. After equilibrium has been reached, the pressure of C is temporarily increased to 1.500 atm. Calculate the partial pressures of A, B, and C after equilibrium has been reestablished.

5. State in which direction the equilibrium will shift to relieve the following stresses on the system.
   a. Increase T.

   b. Increase V.

   c. Add Ar.

   d. Increase the concentration of A.

6. Recall that K is 50.0 at 100°C. At 23°C, K is 137. What is $\Delta H°$ for the forward reaction?

ANSWERS
Exercises:

(E1) $K = \dfrac{(P_{NO})^4 \, (P_{H_2O})^6}{(P_{NH_3})^4 \, (P_{O_2})^5}$

(E2) $K = (P_{NO_2})^2 \, (P_{O_2})^{\frac{1}{2}}$

(E3) 2.0×10^1

(E4) $NH_4Cl \ (s) \ \rightleftharpoons \ NH_3 \ (g) + HCl \ (g) \ ; \quad K = (P_{NH_3}) \, (P_{HCl}) \ ; \quad K = 0.544$

(E5) 0.953

(E6) 0.522

(E7) left to right ($\rightarrow$)

(E8) 2.1 mg

(E9) $P_{H_2} = P_{CO_2} = 0.13$ atm; $\quad P_{CO} = 0.12$ atm; $\quad P_{H_2O} = 0.22$ atm

(E10) a. $\rightarrow$ **b.** $\leftarrow$ **c.** $\leftarrow$ **d.** $\rightarrow$ **e.** $\rightarrow$

(E11) $P_{H_2} = 0.619$ atm; $\quad P_{CO_2} = 0.219$ atm; $\quad P_{CO} = 0.281$ atm; $\quad P_{H_2O} = 0.481$ atm

(E12) 953 K

Self-Test

A. Multiple Choice:

1. b	**2.** b	**3.** d	**4.** b	**5.** c
6. d	**7.** d	**8.** a	**9.** a	**10.** b

B. Circle the correct answer:

1. I   NC   **2.** I   NC   **3.** I   D   **4.** NC   NC   **5.** D   NC

C. Fill in the blanks:

1. 2 atm   **2.** no change   **3.** +   **4.** ←   **5.** 1

D. Problems:

1. $K = \dfrac{(P_C)^2}{(P_A)(P_B)}$

2. 7.41

3. 0.148

4a. Reaction goes to the right

4b. $P_A = P_B = 0.110$ atm; $P_C = 0.780$ atm

4c. $P_A = P_B = 0.190$ atm; $P_C = 1.341$ atm

5a. ←   **b.** no change   **c.** no change   **d.** →

6. −12.0 kJ

13

Acids and Bases

I. **Bronsted-Lowry Acid-Base Model**
  A. Working definitions (from Chapter 4)
    1. Acid
       A substance that when added to water produces hydrogen (H^+) ions (protons)
    2. Base
       A substance that when added to water produces hydroxide (OH^-) ions
  B. Bronsted-Lowry model
    1. An acid is a proton (H^+) donor.
    2. A base is a proton acceptor.
    3. In an acid-base reaction, a proton is transferred from an acid to a base.
    4. Conjugate acid

       Definition:_____

       _____
    5. Conjugate base

       Definition:_____

       _____
    6. Amphiprotic species

       Definition:_____

       _____
    7. H_3O^+ is called a hydronium ion.

II. The Ion Product of Water

   A. Reactions

   1. Showing water's amphiprotic nature

$$H_2O + H_2O \rightleftharpoons H_3O^+ \text{ (aq)} + OH^- \text{ (aq)}$$

   2. Simple ionization

$$H_2O \rightleftharpoons H^+ \text{ (aq)} + OH^- \text{ (aq)}$$

   B. Ion product constant – K_w

   1. Equilibrium constant expression

   2. Value for K_w at 25°C: _____

   3. *Remember:*

   a. K_w is written for the breaking up of the water molecule.

   b. Water (H_2O) is not involved in the equilibrium expression, because it is a pure liquid.

   4. [H^+] for pure water at 25°C is _____ .

   [OH^-] for pure water at 25°C is _____ .

   5. Definitions of acids, bases and neutral solutions based on [H^+]

   a. The solution is acidic if [H^+] is _____ 1×10^{-7} M.

   b. The solution is basic if [H^+] is _____ 1×10^{-7} M.

   c. The solution is neutral if [H^+] is _____ 1×10^{-7} M.

   6. *Exercises*

   a. What is the [H^+] of a sample of lake water with [OH^-] of 4.0×10^{-9}? Is the lake water acidic, basic or neutral?

   Solution: Since $K_w = [H^+][OH^-]$ and $K_w = 1.0 \times 10^{-14}$,

$$[H^+] = \frac{K_w}{[OH^-]} = \frac{1.0 \times 10^{-14}}{4.0 \times 10^{-9}} = 2.5 \times 10^{-6} \text{ M}$$

   Because [H^+] > 1×10^{-7}, the lake water is acidic.

   Remember: The smaller the negative exponent is, the larger the number is. Therefore, acid solutions should have exponents for [H^+] from 0 to –6. Base solutions will have exponents of [H^+] from –8 on.

   b. What is the [H^+] of human saliva if its [OH^-] is 4×10^{-8}? Is human saliva acidic, basic or neutral? **(E1)**

III. pH and pOH
    A. Relationships
        1. Relationship between $[H^+]$ and pH

        2. Relationship between $[OH^-]$ and pOH

        3. Relationship between $[H^+]$ and pH and pOH

    B. Definition of acidic, basic, and neutral solutions based on pH
        1. The solution is acidic if pH is _____ 7.
        2. The solution is basic if pH is _____ 7.
        3. The solution is neutral if pH is _____ 7.
    C. Exercises
        1. Calculate $[OH^-]$ for a solution of baking soda with a pH of 8.50. Is the solution acidic, basic or neutral?
        Solution:
        a. Since the pH is 8.50, which is greater than 7, the solution is basic.
        b. Since we want to calculate $[OH^-]$, we first need to know the pOH. Using the relation between pH and pOH, we get

$$pOH = 14.00 - 8.50 = 5.50$$

We now calculate $[OH^-]$.

$$pOH = -\log[OH^-]$$
$$5.50 = -\log[OH^-]$$
$$[OH^-] = \text{inv} \log(-5.50) = 3.2 \times 10^{-6}\,M$$

Note that you take the inverse log of −5.50 on your calculators and NOT −(inv log 5.50). Note too that it is LOG (to the base 10) and not LN (base e) that you need to use.
Remember: You cannot have a negative number for a concentration.
Reminder: Learn how to use your calculator to determine logs and inverse logs. Read the manual that comes with it or seek help.

2. Calculate the pH of a solution of household ammonia whose OH^- concentration is 7.93×10^{-3}. **(E2)**

D. Strong acids and strong bases
   1. Review from Chapter 4
      a. The strong acids are: _____

_____

      b. The strong bases are: _____

_____

   2. pH of strong acids
      a. All strong acids except H_2SO_4 have the same H^+ concentration as the molarity of the strong acid. This is because these strong acids ionize completely, giving one mole of H^+/mole of strong acid.
      b. To determine the pH of a strong acid (except H_2SO_4), we simply take the log of the molarity of the strong acid and add a minus sign.
      c. *Exercises*
         (1) Calculate $[H^+]$ and the pH of a 0.250 M solution of nitric acid, HNO_3.
         Solution: Since HNO_3 is a strong acid, all the acid ionizes.

$$HNO_3 \text{ (aq)} \quad \rightarrow \quad H^+ \text{ (aq)} + NO_3^- \text{ (aq)}$$

The H^+ concentration is the same as the HNO_3 concentration, which is 0.250 M. The pH therefore is

$$pH = -\log [H^+] = -\log (0.250) = 0.602$$

         (2) What is the molarity of a solution of hydrobromic acid with a pH of 1.78? **(E3)**

   3. pOH and strong bases
      a. Strong bases from Group 1 metals ionize to give _____ mol OH^-/mol of strong base.
      b. Strong bases from Group 2 metals ionize to give _____ mol OH^-/mol of strong base.

c. Keep the facts from (**a**) and (**b**) in mind when you want to calculate the pH of a strong base, given its molarity.

d. *Exercises*

(*1*) Calculate the pH of a solution prepared by dissolving 2.08 g of KOH in enough water to make 500.0 mL of solution.

Solution: We first need to determine the molarity of the base. Since we are given a mass of solute and a volume of solution, we obtain

$$\frac{2.08 \text{ g KOH}}{0.5000 \text{ L solution}} \times \frac{1 \text{ mol KOH}}{56.1 \text{ g KOH}} = 0.0742 \frac{\text{moles}}{\text{L}} = 0.0742 \text{ M}$$

We see that KOH is a strong base from a group 1 metal, K. Thus we get one mole of OH^- for every mole of KOH. In this particular solution, therefore

$$0.0742 \text{ M KOH} \rightarrow 0.0742 \text{ M } OH^-$$

Thus $[OH^-] = 0.0742$ M. We now solve for pOH and pH.

$$pOH = -\log(0.0742) = 1.130$$

$$pH = 14.00 - 1.130 = 12.87$$

(*2*) A student is to dissolve 8.75 g of $Sr(OH)_2$ in enough water to get a pH of 13.35. What should the final volume of the solution be? (**E4**)

E. Measuring pH
  1. Instrument: pH meter
  2. Acid-base indicators

IV. Weak Acids and their Equilibrium Constants
  A. General description
    1. Definition:
      A weak acid is one that only partially ionizes in water to give H_3O^+ and the anion.
    2. *Example*
      HF ionizes in water to give H_3O^+ and F^-. It is a weak acid. Its ionization equation is

$$HF \text{ (aq)} + H_2O \rightleftharpoons H_3O^+ \text{ (aq)} + F^- \text{ (aq)}$$

Note: Use double arrows when writing the ionization equation for a weak acid, because an equilibrium exists between the ionized ions and the unionized molecule. In the case of a strong acid dissociating, only one arrow ($\rightarrow$) is required since the reaction goes virtually to completion.

3. **What are weak acids?**

   There are thousands of weak acids. The easiest way to determine whether an acid is weak is to see whether reaction of the species with water yields an H_3O^+ ion. If it does and it is not one of the six strong acids, then the acid must be weak.

4. **Species that can act as weak acids**
   a. Molecules containing an ionizable hydrogen atom:
      Any molecule that starts with an H is a prime candidate for an acid.
      Example: _____
   b. Cations
      (1) NH_4^+
          The ammonium ion in water yields H_3O^+ and NH_3.
      (2) Transition metal cations
          These are hydrated ions, formed by the bare transition metal ion and water, for example, $Zn(H_2O)_4^{2+}$. In water solutions, these species can transfer a proton from a bonded water molecule to a solvent water molecule. The remaining OH^- remains bonded to the cation, thus reducing the cation charge by 1.

$$Zn(H_2O)_4^{2+} \text{ (aq)} + H_2O \rightleftharpoons Zn(H_2O)_3(OH)^+ \text{ (aq)} + H_3O^+ \text{ (aq)}$$

B. Expression for K_a
   1. The equilibrium expression for the ionization of a weak acid follows the general rules given in Chapter 12 of your text.
   2. The ions and molecules in the equilibrium expression are enclosed in brackets. This means that the units for these species are expressed in moles/liter (M).
   3. The concentration of the unionized weak acid at equilibrium $[HB]_{eq}$ is equal to the concentration of the weak acid before ionization $[HB]_o$ minus the concentration of one of the ions. Thus

$$[HB]_{eq} = [HB]_o - [H^+] \quad \text{or} \quad [HB]_{eq} = [HB]_o - [B^-]$$

4. pKa
   Mathematical relationship of pKa to K_a:

5. *Exercises*

    a. Consider acetic acid, $HC_2H_3O_2$.

        (1) Using the Bronsted-Lowry model, write an equation to show why it behaves as a weak acid in water.

        Solution: The equation is

$$HC_2H_3O_2 \text{ (aq)} + H_2O \rightleftharpoons C_2H_3O_2^- \text{ (aq)} + H_3O^+ \text{ (aq)}$$

This equation shows the proton transfer to a solvent water molecule.

        (2) Write the equation for the ionization of the acetic acid molecule.

        Solution: For this type of equation, we simply peel off the hydrogen that starts the molecule. The equation is

$$HC_2H_3O_2 \text{ (aq)} \rightleftharpoons C_2H_3O_2^- \text{ (aq)} + H^+ \text{ (aq)}$$

This is the equation that we will use more frequently.

Remember that in the ionization reactions of a weak acid, the whole molecule or ion is on the left side of the equation. The right side will have *one* H^+ and the rest of the molecule or ion. The charge of the rest of the molecule or ion on the right must be one less than the charge of the molecule or ion on the left. The rest of the ion or molecule on the right is the conjugate base of the weak acid. In our exercise, the conjugate base of acetic acid ($HC_2H_3O_2$) is the acetate ion, ($C_2H_3O_2^-$).

        (3) Write the equilibrium constant expression, K_a, for acetic acid in water.

        Solution: For the K_a expression, it is more convenient to use the simple ionization equation instead of the one that shows the proton transfer to the solvent water molecule. For acetic acid it is

$$K_a = \frac{[H^+][C_2H_3O_2^-]}{[HC_2H_3O_2]}$$

    Note: K_a always expresses *ionization*. Hence H^+ is always in the numerator.

    b. Consider nitrous acid, HNO_2, in water.

        (1) Using the Bronsted-Lowry model, write an equation to show why it behaves as a weak acid in water.

        (2) Write an equation for the simple ionization of HNO_2 in water.

        (3) What is the conjugate base of HNO_2?

(4) Write the expression for its equilibrium constant, K_a. **(E5)**

C. Qualitative interpretation of K_a
  1. The smaller the acid equilibrium constant is, the weaker the acid is.
  2. When comparing two weak acids with the same concentration, the acid with the smaller equilibrium constant
     a. is weaker.
     b. has a smaller $[H^+]$.
     c. has a higher pH.
     d. has a larger $[HB]$.
     e. has a larger pKa.
D. Determination of K_a
  1. Given a known amount of weak acid in a given volume, as well as the pH of the resulting solution
     This method gives you the original concentration of the weak acid and $[H^+]$.
     Exercises
     a. A solution of $HClO_2$ is prepared by dissolving 1.369 g of $HClO_2$ in enough water to make 100.0 mL of solution. The pH of the resulting solution is 1.36.
        (1) Write the reaction for the ionization of $HClO_2$.
           Solution:
$$HClO_2 \text{ (aq)} \rightleftharpoons H^+ \text{ (aq)} + ClO_2^- \text{ (aq)}$$

        (2) Write the expression for K_a.
           Solution:
$$K_a = \frac{[H^+][ClO_2^-]}{[HClO_2]}$$

        (3) Calculate K_a.
           Solution: To calculate K_a, we need to know the equilibrium concentrations of all the species. It will be helpful if you make a table just as you did in Chapter 12.

	$HClO_2$	$\rightleftharpoons$	H^+	$+$	ClO_2^-
[ ]$_0$					
Δ[ ]					
[ ]$_{eq}$					

           In this table [ ]$_0$ stands for original concentration, Δ[ ] stands for change in the concentration and [ ]$_{eq}$ stands for equilibrium concentration. All concentrations are expressed in molarity (M).

Since only $HClO_2$ was present originally, you only have to calculate the original concentration of $HClO_2$ in moles/liter. The gram molar mass of $HClO_2$ is 68.46 g/mol. The number of moles of $HClO_2$ therefore is

$$1.369 \, g \times \frac{1 \, mol}{68.46 \, g} = 0.02000 \, mol$$

Its concentration in moles/liter is

$$\frac{0.02000 \, mol}{0.1000 \, L} = 0.2000 \, M$$

Your table now looks like this:

	$HClO_2$	$\rightleftharpoons$	H^+	+	ClO_2^-
[ ]$_o$	0.2000		0		0
Δ[ ]					
[ ]$_{eq}$					

The fact that you are given the pH of the solution means that you can get $[H^+]$ or H^+ concentration at equilibrium ($[H^+]_{eq}$). Calculating for $[H^+]$ from pH, we have

$$pH = -\log[H^+]$$
$$1.36 = -\log[H^+]$$

Taking the inverse log of both sides, we obtain

$$[H^+] = \text{inv} \log(-1.36) = 0.044$$

In addition to writing this information on the table, we can also write the pluses (for products) and the minuses (for reactants). Your table will now look like this:

	$HClO_2$	$\rightleftharpoons$	H^+	+	ClO_2^-
[ ]$_o$	0.2000		0		0
Δ[ ]	−		+		+
[ ]$_{eq}$			0.044		

Doing the arithmetic for the H^+ column, we see that $\Delta[H^+]$ is 0.044. Since one mole of $HClO_2$ ionizes into one mole of H^+ and one mole of ClO_2^-, $\Delta[H^+] = \Delta[ClO_2^-] = \Delta[HClO_2]$. Your table now looks like this:

	$HClO_2$	$\rightleftharpoons$	H^+	$+$	ClO_2^-
$[\ ]_o$	0.2000		0		0
$\Delta[\ ]$	-0.044		$+0.044$		$+0.044$
$[\ ]_{eq}$			0.044		

Finally, doing the arithmetic for each column, we get a table like this:

	$HClO_2$	$\rightleftharpoons$	H^+	$+$	ClO_2^-
$[\ ]_o$	0.2000		0		0
$\Delta[\ ]$	-0.044		$+0.044$		$+0.044$
$[\ ]_{eq}$	0.156		0.044		0.044

Now we can calculate K_a by using the values in the last row to plug into the equilibrium expression

$$K_a = \frac{[H^+][ClO_2^-]}{[HClO_2]} = \frac{(0.044)(0.044)}{0.156} = 0.012$$

b. A one-liter solution of nitrous acid (HNO_2) contains 11.76 g of the weak acid. The solution has a pH of 1.955. Calculate K_a for HNO_2. **(E6)**

2. Given the percent ionization of the acid and its original concentration
    Exercises
    a. A 0.300 M aqueous solution of acid ($HC_7H_5O_2$) is 1.47% ionized.
        (1) Write the equation for the ionization of benzoic acid.
            Solution:

$$HC_7H_5O_2\ (aq) \rightleftharpoons H^+\ (aq) + C_7H_5O_2^-\ (aq)$$

        (2) Write the K_a expression for the ionization.
            Solution:

$$K_a = \frac{[H^+][C_7H_5O_2^-]}{[HC_7H_5O_2]}$$

(3) Calculate the K_a for the ionization.

Solution: We make a table and put in the given data for the first row. It will look like this

	$HC_7H_5O_2$	$\rightleftharpoons$	H^+	+	$C_7H_5O_2^-$
[ ]$_o$	0.3000		0		0
Δ[ ]	−		+		+
[ ]$_{eq}$					

1.47% ionization means that 1.47% of $HC_7H_5O_2$ has been ionized, that is, has been *changed* to ions. Thus $\Delta[HC_7H_5O_2]$ is

$$0.300 \times 0.0147 = 0.00441$$

Since the reaction shows that for every mole of $HC_7H_5O_2$ that ionizes, one mole of H^+ and one mole of $C_7H_5O_2^-$ are formed, then

$$\Delta[HC_7H_5O_2] = \Delta[H^+] = \Delta[C_7H_5O_2^-] = 0.00441$$

We use this information to complete the second row of the table, which will now look like this:

	$HC_7H_5O_2$	$\rightleftharpoons$	H^+	+	$C_7H_5O_2^-$
[ ]$_o$	0.3000		0		0
Δ[ ]	− 0.00441		+ 0.00441		+ 0.00441
[ ]$_{eq}$					

We can now complete the third row of the table:

	$HC_7H_5O_2$	$\rightleftharpoons$	H^+	+	$C_7H_5O_2^-$
[ ]$_o$	0.3000		0		0
Δ[ ]	− 0.00441		+ 0.00441		+ 0.00441
[ ]$_{eq}$	0.2956		0.00441		0.00441

Solving for K_a, using the data in the third row of the table to plug into the K_a expression, we get

$$K_a = \frac{(0.00441)(0.00441)}{0.2956} = 6.58 \times 10^{-5}$$

b. Consider the ionization of the amphiprotic anion, $HC_2O_4^-$, in aqueous solution.

$$HC_2O_4^- \text{ (aq)} \rightleftharpoons H^+ \text{ (aq)} + C_2O_4^{2-} \text{ (aq)}$$

A 0.0100 M solution of the anion is 7.35% ionized. Calculate K_a for this ionization. **(E7)**

E. Calculation of [H⁺] in a water solution of a weak acid

This is analogous to determining the extent of the reaction in Chapter 12.

Exercises

1. Phenol (HOC_6H_5) has an acid equilibrium constant of 1.6×10^{-10}. Determine [H⁺] in a solution prepared by dissolving 0.500 mol of phenol to form 5.00 L of solution.

Solution: The equation for the ionization of phenol is

$$HOC_6H_5 \text{ (aq)} \rightleftharpoons H^+ \text{ (aq)} + OC_6H_5^- \text{ (aq)}$$

Starting our table, we get

	HOC_6H_5	$\rightleftharpoons$	H^+	+	$OC_6H_5^-$
[ ]₀	0.100		0		0
Δ[ ]	−		+		+
[ ]eq					

We need to calculate [H⁺]eq, so we designate [H⁺]eq as x. Δ[H⁺] is therefore x too. And again, since 1 mole of phenol ionizes into 1 mole of H⁺ and 1 mole of $OC_6H_5^-$

$$\Delta [H^+] = \Delta [OC_6H_5^-] = \Delta [HOC_6H_5] = x$$

Our table now looks like this:

	HOC_6H_5	$\rightleftharpoons$	H^+	+	$OC_6H_5^-$
$[\ ]_o$	0.100		0		0
$\Delta[\ ]$	$-x$		$+x$		$+x$
$[\ ]_{eq}$			x		

Filling in the rest of the table, we get

	HOC_6H_5	$\rightleftharpoons$	H^+	+	$OC_6H_5^-$
$[\ ]_o$	0.100		0		0
$\Delta[\ ]$	$-x$		$+x$		$+x$
$[\ ]_{eq}$	$0.100 - x$		x		x

Substituting into the equilibrium expression

$$K_a = \frac{[H^+][OC_6H_5^-]}{[HOC_6H_5]}$$

we have

$$1.6 \times 10^{-10} = \frac{(x)\,(x)}{0.100 - x}$$

Most weak acids have very small equilibrium constants. We therefore may assume that the amount ionized is so much smaller than the original concentration that it is negligible when compared to the original concentration. In the preceding equation, then, we can assume $0.100 - x \approx 0.100$. You must always write out your assumption and test later whether it is valid.

We now write: Assume $x \ll 0.100$. Thus $0.100 - x \approx 0.100$ and

$$1.6 \times 10^{-10} \approx \frac{x^2}{0.100}$$

$$x^2 = 0.16 \times 10^{-10}$$

$$x = 4.0 \times 10^{-6}$$

Since $[H^+] = x$, then $[H^+] = 4.0 \times 10^{-6}$ M.

We now test our assumption. For the assumption to be valid, the ionization must be less than 5%.

$$\% \text{ ionization} = \frac{x}{[HB]_o} \times 100$$

$$= \frac{4.0 \times 10^{-6}}{0.100} \times 100 = 0.0040\%$$

Since $0.0040\% < 5\%$, the assumption is valid. If the percent ionization is greater than 5%, then go back and recalculate x using the quadratic equation. The method of successive approximations described in the text can also be tried. Use the method at which you are most adept.

2. Consider the hydrogen sulfite ion (HSO_3^-). The K_a for the following ionization is 6.0×10^{-8}.

$$HSO_3^- \ (aq) \ \rightleftharpoons \ H^+ \ (aq) \ + \ SO_3^{2-} \ (aq)$$

Calculate $[H^+]$ for a 5.00-L solution containing 0.394 mol of HSO_3^-. **(E8)**

F. Polyprotic Acids

1. Definition of polyprotic: _____

   _____

2. Weak acids with more than one ionizable H^+, H_3PO_4 for instance, ionize step-wise; that is, they peel off one H^+ at a time. Each step has a different K_a value. The K_a value decreases with each successive step. The steps for the ionization of H_3PO_4 are:

$$H_3PO_4 \ (aq) \ \rightleftharpoons \ H^+ \ (aq) \ + \ H_2PO_4^- \ (aq)$$
$$H_2PO_4^- \ (aq) \ \rightleftharpoons \ H^+ \ (aq) \ + \ HPO_4^{2-} \ (aq)$$
$$HPO_4^{2-} \ (aq) \ \rightleftharpoons \ H^+ \ (aq) \ + \ PO_4^{3-} \ (aq)$$

Note that the conjugate base of the first step becomes the weak acid of the second step. These are amphiprotic species.

Write the equations for the stepwise ionization of H_2CO_3. **(E9)**

3. Essentially all the hydrogen ions in the solution of a polyprotic acid come from the first ionization of the acid.

V. Weak Bases

A. Species that act as weak bases
  1. Molecules
     a. Ammonia — NH_3
     b. Amines, which can be represented generally by R_3N where R can be H, CH_3, etc
  2. Anions
     Any anion that is the conjugate of a weak acid

     Example: _____

B. Equilibrium system of a weak base (B^-) in water.
  1. The equation for the reaction is

$$\text{weak base} + H_2O \rightleftharpoons \text{conjugate acid} + OH^- \text{(aq)}$$

  or

$$B^- \text{(aq)} + H_2O \rightleftharpoons HB \text{(aq)} + OH^- \text{(aq)}$$

For example, ammonia (NH_3) is a weak base. When it ionizes in water, the equation for the ionization is

$$NH_3 \text{(aq)} + H_2O \rightleftharpoons NH_4^+ \text{(aq)} + OH^- \text{(aq)}$$

When a weak base ionizes in water, the products always are OH^- and the weak base with one more H attached to it. The "weak base with one more H attached to it" is called the conjugate acid of that base. To get the charge of the conjugate acid, take the charge of the weak base and add 1.

  2. To show the proton transfer in a weak base, water must always be included in the reaction. There is no other simple ionization equation for a weak base as there is for a weak acid.

  3. K_b is called the base equilibrium constant. Its equilibrium expression for the general equation for the ionization of a weak base shown above is

$$K_b = \frac{[OH^-][\text{conjugate acid}]}{[\text{weak base}]}$$

  or

$$K_b = \frac{[OH^-][HB]}{[B^-]}$$

As in Chapter 12, pure water is not included in the equilibrium expression. For the weak base $C_2H_3O_2^-$, the ionization equation is

$$C_2H_3O_2^- \text{(aq)} + H_2O \rightleftharpoons HC_2H_3O_2 \text{(aq)} + OH^- \text{(aq)}$$

The equilibrium expression is

$$K_b = \frac{[OH^-][HC_2H_3O_2]}{[C_2H_3O_2^-]}$$

Remember that the conjugate base of a weak acid is a weak base.

C. Characteristics of K_b
1. To have an equilibrium expression where K_b is used, hydroxide ions must be on the right side (product side) of the equation.
2. The larger K_b is, the stronger the base is.
3. When comparing two bases with the same molarity, the base with the larger K_b will have the higher OH^- concentration, $[OH^-]$, and the higher pH.

D. Relation between K_a and K_b
1. $K_a \times K_b = 1 \times 10^{-14}$
2. Students sometimes have a difficult time looking up K_a's and K_b's in tables and deciding which one to use. In general:
  a. If H^+ is on the product side, use K_a. You should look for the K_a of the weak acid that is on the reactant side of the equation.
  b. If OH^- is on the product side, use K_b. You should look for the K_b of the weak base that is on the reactant side of the equation.
  c. For example:

$$NH_3 \text{ (aq)} + H_2O \rightleftharpoons NH_4^+ \text{ (aq)} + OH^- \text{ (aq)}$$

   requires the K_b expression. Look up K_b for NH_3.
   On the other hand

$$NH_4^+ \text{ (aq)} \rightleftharpoons NH_3 \text{ (aq)} + H^+ \text{ (aq)}$$

   requires the K_a expression. Look up K_a for NH_4^+.

E. Determination of $[OH^-]$ in an aqueous solution of a weak base.
The determination of $[OH^-]$ parallels the calculation of $[H^+]$ from the ionization of a weak acid. You should construct a table and make similar assumptions.

Exercises

1. Hydroxylamine ($HONH_2$) ionizes in water according to the equation

$$HONH_2 \text{ (aq)} + H_2O \rightleftharpoons OH^- \text{ (aq)} + HONH_3^+ \text{ (aq)}$$

The base equilibrium constant, K_b, for this ionization is 9.1×10^{-9}. Calculate $[OH^-]$ and the pH of a one-liter solution containing 0.20 M hydroxylamine.

Solution: To calculate $[OH^-]$, we again construct a table.

	$HONH_2$	+	H_2O	$\rightleftharpoons$	OH^-	+	$HONH_3^+$
$[\ ]_0$	0.20				0		0
$\Delta[\ ]$	$-x$				$+x$		$+x$
$[\ ]_{eq}$	$0.20-x$				x		x

Notice that we made no entries in the H_2O column. That is because H_2O does not appear in the K_b expression.

We now substitute the entries of the table's third row into the equilibrium expression

$$K_b = \frac{[OH^-][HONH_3^+]}{[HONH_2]}$$

and get

$$9.1 \times 10^{-9} = \frac{(x)(x)}{0.20 - x}$$

Assuming $x << 0.20$, we have

$$9.1 \times 10^{-9} = \frac{(x)^2}{0.20}$$

Solving for x we obtain

$$x = 4.3 \times 10^{-5} = [OH^-]$$

Checking our assumption, we see that the ionization is less than 5%, so our assumption is valid. We now calculate pH by first calculating pOH.

$$pOH = -\log(4.3 \times 10^{-5}) = 4.37$$

We next use the relationship $pH + pOH = 14.00$ to calculate pH.

$$pH = 14.00 - 4.37 = 9.63$$

2. Determine $[OH^-]$ and pH of a 0.100 M solution of methylamine, CH_3NH_2, with $K_b = 4.2 \times 10^{-4}$. **(E10)**

VI. Acid-Base Properties of Salt Solutions

  A. Definition of a salt:

_____

  B. Determination of the acidity or basicity of a salt solution in water

    1. Table 13.5 of your text lists the most common spectator, acidic, and basic cations and anions. If you can remember these, then you should have no trouble determining the acidity or basicity of a salt solution.

    2. Another simple way to determine the acidity or basicity of the ions is to go through the following steps.

      a. Split the salt into cation and anion.

      b. Ask: Is the cation a metal of Group 1 or 2?

        (1) yes — The cation is a spectator ion.

        (2) no — The cation is acidic.

        Remember: there are no basic cations.

      c. Ask: Is the anion strong acid related? (i.e., Is it Cl^-, Br^-, I^-, ClO_4^-, NO_3^-, or SO_4^{2-}?)

        (1) yes — The anion is a spectator ion.

        (2) no — Is the anion amphiprotic?

          (a) no — The anion is basic.

          (b) yes — Do the following:

            — Write out the equation for the ionization of the species as a weak acid. Look up K_a for the weak acid.

            — Write out the equation for the ionization of the species as a weak base. Look up K_b for the weak base.

            — If $K_a > K_b$, the species is acidic. If not, then it is basic.

    3. Determining the acidity of the salt once the acidity of the individual ions is known

      a. Salt solutions are neutral if both ions are spectator ions.

      b. Salt solutions are acidic if one ion is a spectator ion and the other is acidic.

      c. Salt solutions are basic if one ion is a spectator ion and the other is basic.

      d. The acidity or basicity of a salt solution made up of one acidic ion and one basic ion cannot be determined without further information. The information needed is the K_a of the acidic ion and the K_b of the basic ion. If $K_a > K_b$, then the salt is acidic; otherwise it is basic.

    _4._ _Exercises_

      _a._ Determine whether an aqueous solution of $KC_2H_3O_2$ is acidic, basic, or neutral.

        Solution:

        _(1)_ Using the method in your text, we split the salt into its cation K^+, which according to the table is a spectator ion, and its anion $C_2H_3O_2^-$, which is basic. Thus the solution is basic.

        _(2)_ We use the method of asking questions outlined earlier:

          _(a)_ We split the salt into its cation, K^+, and its anion, $C_2H_3O_2^-$.

          _(b)_ Is the cation a metal of Group 1 or Group 2? Yes. The cation is a spectator ion.

(*c*) Is the anion strong-acid related? No.
Is it amphiprotic? No.
The anion is basic.

Since we have a salt with a spectator cation and a basic anion, an aqueous solution of the salt is basic.

b. Determine whether the following aqueous solutions are acidic, basic, or neutral.
$Cu(NO_3)_2$; $KClO_4$; NaH_2PO_4; LiF; $(NH_4)_2CO_3$ **(E11)**

5. Determining the pH of a salt solution.

Exercises

a. Determine the pH of a 0.100 M aqueous solution of NaCN. The K_a for HCN is 5.8×10^{-10}.

Solution: In aqueous solutions, NaCN ionizes completely into Na^+ and CN^-. In turn, the cyanide ions ionize in water according to the following equation

$$CN^- \text{ (aq)} + H_2O \rightleftharpoons HCN \text{ (aq)} + OH^- \text{ (aq)}$$

Since the Na^+ ions are spectator ions, the equilibrium expression for the solution is

$$K_b = \frac{[OH^-][HCN]}{[CN^-]}$$

We need a value for K_b. Since we have K_a for HCN, we can calculate the value of K_b for CN^-.

$$K_b = \frac{1.0 \times 10^{-14}}{K_a} = \frac{1.0 \times 10^{-14}}{5.8 \times 10^{-10}} = 1.7 \times 10^{-5}$$

Constructing a table for filling in the information we have, we get

	CN^-	+	H_2O	$\rightleftharpoons$	OH^-	+	HCN
$[\ ]_o$	0.100				0		0
$\Delta[\ ]$	$-x$				$+x$		$+x$
$[\ ]_{eq}$	$0.100 - x$				x		x

Assuming $x \ll 0.1$, we get

$$1.7 \times 10^{-5} = \frac{x^2}{0.100}$$

Solving for x, we have

$$x = [OH^-] = 1.3 \times 10^{-3}$$

$$pOH = 2.88 \qquad pH = 11.12$$

b. Determine the pH of a solution prepared by dissolving 4.00 g of NaF in enough water to make 500.0 mL of solution. The ionization constant for HF is 6.9×10^{-4}. **(E12)**

SELF-TEST
A. Multiple Choice:

1. Which of the following will make a solution basic?

     (1) NaCl   **(2)** Na_3PO_4 **(3)** NH_3 **(4)** NH_4Cl **(5)** HCN   **(6)** KCN

     a. (1),(2),(3)   **b.** (2),(3),(4)   **c.** (2),(3),(6)   **d.** (3),(4),(6)

2. When 1×10^{-5} mole of NaOH is added to enough water to make 1.0 L of solution, the resulting solution has a pH of
     a. 5   **b.** 6   **c.** 7   **d.** 8   **e.** 9

3. Vinegar generally contains 5% acetic acid. We would expect the pH of vinegar to be approximately
     a. 0   **b.** 3   **c.** 7   **d.** 9   **e.** 12

4. Which pair below consists of a Bronsted acid and a Bronsted base?
     a. HI, NH_4^+ **b.** SO_3^-, $H_2PO_4^-$ **c.** NH_3, PO_4^{3-} **d.** H_2CO_3, H_3O^+

5. As the hydrogen ion concentration decreases in an aqueous solution

     — hydroxide ion concentration increases.
     — pH increases.
     — the solution becomes less acidic.
     — K_w increases.

     The number of correct choices is
     a. 0   **b.** 1   **c.** 2   **d.** 3   **e.** 4

6. For the acetic acid/acetate ion system, the following ionization constants are defined:

$$HA\ (aq)\ \rightleftharpoons\ H^+\ (aq)\ +\ A^-\ (aq) \qquad K_a$$
$$A^-\ (aq)\ +\ H_2O\ \rightleftharpoons\ HA\ (aq)\ +\ OH^-\ (aq) \qquad K_b$$

     The true expressions are

     (1) $K_a + K_b = 1$ **(2)** $K_a \times K_b = K_w$ **(3)** $K_a/K_b = 1$ **(4)** $K_a + K_b = K_w$

     a. (1)   **b.** (1),(2)   **c.** (3),(4)   **d.** (2)   **e.** none of these

7. The pH of a 0.1 M solution of H_2SO_4 is

     a. between 0 and 1   **b.** 1   **c.** between 1 and 2
     d. 2   **e.** None of those

8. Consider solutions of the following salts

     (1) $CsClO_4$ **(2)** $Cr(NO_3)_3$ **(3)** CH_3NH_2

     Listed in order of increasing pH, the correct order is
     a. (1),(2),(3)   **b.** (2),(1),(3)   **c.** (1),(3),(2)   **d.** (3),(2),(1)

9. Given the following K_a values

HC$_2$H$_3$O$_2$	1.8×10^{-5}	HBrO	5.8×10^{-10}
HCN	5.8×10^{-10}	HF	6.9×10^{-4}
HOCl	2.8×10^{-8}		

Which of the following is the weakest base?

a. $C_2H_3O_2^-$ **b.** BrO^- **c.** CN^- **d.** F^- **e.** OCl^-

10. Sulfurous acid is a diprotic acid. Its ionization occurs in the following steps:

$$H_2SO_3 \text{ (aq)} \rightleftharpoons H^+ \text{ (aq)} + HSO_3^- \text{ (aq)} \qquad K_{a_1} = 1.3 \times 10^{-2}$$
$$HSO_3^- \text{ (aq)} \rightleftharpoons H^+ \text{ (aq)} + SO_3^{2-} \text{ (aq)} \qquad K_{a_2} = 6.3 \times 10^{-8}$$

In a 0.10 M H_2SO_3 solution whose pH has been adjusted to 10, which of the following species would have the highest concentration?

a. SO_3^{2-} **b.** HSO_3^- **c.** H^+ **d.** H_2SO_3 **e.** H_2O

B. Fill in the Blanks

Consider aqueous solutions of the following salts. On the space provided, write **A** if the solution is *acidic* , **B** if the solution is *basic* or **N** if the solution is *neutral.*

_____ 1. FeCl$_3$

_____ 2. NaI and KNO$_3$

_____ 3. Mg(OH)$_2$ and CaCO$_3$

_____ 4. CsClO$_4$ and KNO$_2$

_____ 5. NH$_4$Br and KBr

C. Problems:

Consider two solutions. One solution is 0.1385 M Ba(OH)$_2$. The other is 0.2050 M HBr.
1. Calculate the pH of the HBr solution.

2. Calculate the pH of the Ba(OH)$_2$ solution.

3. If 50.00 mL of the $Ba(OH)_2$ is mixed with 30.00 mL of the HBr solution, what is the pH of the resulting solution?

Consider propionic acid, $HC_3H_5O_2$.

4. Write a balanced equation for the reaction that takes place when it is dissolved in water.

5. Write the K_a expression for this reaction.

6. Calculate K_a for propionic acid if a 500.0-mL solution of propionic acid containing 9.250 g of the acid has a pH of 2.73.

7. A different solution of propionic acid is prepared by dissolving 1.5000 g in enough water to make 2.000 L of solution. Using the K_a calculated in (6), what is the pH of the resulting solution?

8. Is sodium propionate, $NaC_3H_5O_2$ acidic, basic, or neutral? Write an equation to prove your answer.

9. Calculate K_b for $C_3H_5O_2^-$.

10. A 750.0-ml solution contains 20.00 g of $NaC_3H_5O_2$. What is the pH of this solution?

ANSWERS
Exercises:
(E1) 2.5×10^{-7}; slightly acidic   **(E2)** 11.899
(E3) 0.017 M   **(E4)** 0.65 L
(E5) (1) HNO_2 (aq) + H_2O $\rightleftharpoons$ H_3O^+ (aq) + NO_2^- (aq)
         (2) HNO_2 (aq) $\rightleftharpoons$ H^+ (aq) + NO_2^- (aq)
         (3) NO_2^- is the conjugate base.
         (4) $K_a = \dfrac{[H^+][NO_2^-]}{[HNO_2]}$

(E6) 5.15×10^{-4} **(E7)** 5.83×10^{-5}
(E8) 6.9×10^{-5} M
(E9) H_2CO_3 (aq) $\rightleftharpoons$ H^+ (aq) + HCO_3^- (aq); HCO_3^- (aq) $\rightleftharpoons$ H^+ (aq) + CO_3^{2-} (aq)
(E10) $[OH^-] = 6.3 \times 10^{-3}$; pH = 11.80
(E11) acidic, neutral, acidic, basic, basic   **(E12)** 8.22

Self-Test
A. Multiple Choice:
  1. c   **2.** e   **3.** b   **4.** b   **5.** d
  6. d   **7.** a   **8.** b   **9.** d   **10.** a

B. Fill in the blanks
  1. A   **2.** N   **3.** B   **4.** B   **5.** A

C. Problems:

1. 0.6882

2. 13.44

3. 12.98

4. $HC_3H_5O_2$ (aq) $\rightleftharpoons$ H^+ (aq) + $C_3H_5O_2^-$ (aq)

5. $K_a = \dfrac{[H^+][C_3H_5O_2^-]}{[HC_3H_5O_2]}$

6. 1.4×10^{-5}

7. 3.42

8. basic; $C_3H_5O_2^-$ (aq) + H_2O $\rightleftharpoons$ $HC_3H_5O_2$ (aq) + OH^- (aq)

9. 7.1×10^{-10}

10. 9.15

Equilibria in Acid-Base Solutions

I. Buffers

A. Definition: _____

_____

B. Characteristics of a buffer
  1. It contains two species — one that can react with H^+, and the other with OH^-.
  2. $[HB]_{eq} \approx [HB]_o$
     $[B^-]_{eq} \approx [B^-]_o$
  3. The pH of a buffer is independent of volume. Thus

$$[H^+] = K_a \frac{[HB]}{[B^-]} = K_a \frac{(\text{moles HB})}{(\text{moles B}^-)}$$

  4. Henderson-Hasselbalch equation

$$pH = pK_a + \log \frac{[B^-]}{[HB]}$$

C. Determination of $[H^+]$ (pH)
  Exercises
  1. Calculate the pH of a solution with a volume of 1.000 L that contains 2.70 g of HCN and 2.45 g of NaCN. K_a for HCN is 5.8×10^{-10}.
     Solution: There are really three equilibrium systems involved in every buffer. In this buffer, they are

$$HCN\ (aq) \rightleftharpoons H^+\ (aq) + CN^-\ (aq)$$
$$H_2O \rightleftharpoons H^+\ (aq) + OH^-\ (aq)$$
$$CN^-\ (aq) + H_2O \rightleftharpoons HCN\ (aq) + OH^-\ (aq)$$

We always choose the equilibrium system for the ionization of the weak acid. That gives $[H^+]$ directly, and hence pH. We can ignore the ionization of water since the concentrations of H^+ and OH^- are negligible compared to the same concentrations in the buffer system. We therefore write the ionization constant expression for HCN in terms of $[H^+]$:

$$[H^+] = K_a \times \frac{[HCN]}{[CN^-]}$$

$$= 5.8 \times 10^{-10} \times \frac{(mol\ HCN)}{(mol\ CN^-)}$$

$$\text{moles HCN} = 2.70\,g \times \frac{1\,mol}{27.03\,g} = 0.100$$

$$\text{moles } CN^- = \text{moles NaCN} = 2.45\,g \times \frac{1\,mol}{49.0\,g} = 0.0500$$

We now substitute into the ionization constant equation and calculate $[H^+]$.

$$[H^+] = 5.8 \times 10^{-10} \times \frac{0.100\,mol\ HCN}{0.0500\,mol\ CN^-} = 1.2 \times 10^{-9}\,M$$

The pH of the buffer is 8.94.

2. Calculate the pH of a solution that is 0.100 M in both H_2CO_3 and $NaHCO_3$. The K_a for the ionization of H_2CO_3 to H^+ and HCO_3^- is 4.4×10^{-7}. **(E1)**

D. Choosing a buffer system

   1. When required to prepare a buffer solution with a particular pH, you are really being asked to determine the ratio of the concentration of the weak acid [HB] to the concentration of its conjugate base $[B^-]$. Follow the steps outlined below to help you in this process.

   a. Calculate $[H^+]$ from the given pH.

   b. Rearrange the ionization constant expression for the weak acid to get the weak acid–conjugate base ratio expression. Substitute pertinent information and calculate to get a numerical ratio.

   c. Assume the moles of conjugate base to be 1.00, and determine the mass of the salt of that conjugate base accordingly.

d. The number of moles of weak acid is the numerical ratio obtained in (b). Determine the mass of weak acid corresponding to that number. You now have the "recipe" for the buffer.

2. *Exercises*

a. Suppose that you want to prepare an $H_2PO_4^-$ – HPO_4^{2-} buffer with a pH of 7.00. Taking the K_a of $H_2PO_4^-$ ($H_2PO_4^- \rightleftharpoons H^+$ (aq) + HPO_4^{2-} (aq)) to be 6.2×10^{-8}, how many grams of NaH_2PO_4 and Na_2HPO_4 should you add to water to make this buffer?

Solution We follow the steps outlined above.

(1) $[H^+]$ = inv log(-7.00) = 1.0×10^{-7} M

(2) We use the ionization constant expression of the weak acid to calculate the weak acid – conjugate base ratio.

$$K_a = \frac{[H^+]\,[HPO_4^{2-}]}{[H_2PO_4^-]}$$

Rearranging and substituting, we get

$$\frac{[H_2PO_4^-]}{[HPO_4^{2-}]} = \frac{[H^+]}{K_a} = \frac{1.0 \times 10^{-7}}{6.2 \times 10^{-8}} = 1.6$$

This means that for every mole of weak base (HPO_4^{2-}), we need 1.6 moles of weak acid ($H_2PO_4^-$).

(3) We now calculate the mass of Na_2HPO_4. Since we need one mole of HPO_4^{2-}, we also need one mole of Na_2HPO_4. Thus

$$\text{mass } Na_2HPO_4 = 1.0 \text{ mol} \times \frac{142\,g}{1\,mol} = 1.4 \times 10^2 \text{ g}$$

(4) Using the same argument as used in (3), we calculate the mass of NaH_2PO_4 required. We see that we need 1.6 moles of weak acid $H_2PO_4^-$, so we also need 1.6 mol of NaH_2PO_4. The mass needed is therefore

$$\text{mass } NaH_2PO_4 = 1.6 \text{ mol} \times \frac{120\,g}{1\,mol} = 1.9 \times 10^2 \text{ g}$$

Therefore, to get a buffer of pH 7.0, dissolve 1.4×10^2 g of Na_2HPO_4 and 1.9×10^2 g of NaH_2PO_4. It doesn't matter how much water you use because the pH of a buffer is not affected by the volume of the solution.

b. How many grams of sodium formate, $NaCHO_2$, must be added to 1.00 L of 0.250 M formic acid, $HCHO_2$, to obtain a buffer with pH 4.00? **(E2)**

E. Adding H^+ or OH^- to a buffer
  1. Adding H^+
    a. Recall that a buffer has one species that can react with H^+ ions. This species is the conjugate base B^-. Thus when H^+ is added to a buffer, the following reaction occurs.

$$H^+ \text{ (aq)} + B^- \text{ (aq)} \rightleftharpoons HB \text{ (aq)}$$

    b. Addition of H^+ does two things to the buffer:
      (1) It makes more acid. The concentration of the weak acid in the buffer system after the addition of H^+ is

$$\text{total moles HB} = (\text{moles HB})_0 + \text{moles } H^+ \text{ added}$$

      (2) It uses up conjugate base. The reaction above shows that for every mole of H^+ added, an equal amount of weak base reacts. Therefore, the concentration of weak base (or conjugate base) after H^+ addition is

$$\text{total moles } B^- = (\text{moles } B^-)_0 - \text{moles } H^+ \text{ added}$$

  2. Adding OH^-
    a. A buffer also has a species that can react with OH^- ions. This species is the weak acid. The reaction is

$$HB \text{ (aq)} + OH^- \text{ (aq)} \rightleftharpoons B^- \text{ (aq)} + H_2O$$

    b. Addition of OH^- ion does two things to a buffer:
      (1) It makes more base. Each mole of OH^- added makes an equivalent amount of weak base B^-. Thus

$$\text{total moles } B^- = (\text{moles } B^-)_0 + \text{moles } OH^- \text{ added}$$

      (2) It uses up weak acid. Each mole of OH^- added uses up an equivalent amount of weak acid.

$$\text{total moles HB} = (\text{moles HB})_0 - \text{moles } OH^- \text{ added}$$

F. Buffer limitations

1. The buffer can "absorb" the large pH changes that adding strong acid or strong base imposes on a system as long as it has enough of the species to react with H^+ or OH^-. A buffer that has 0.100 mol HB and 0.100 mol B^-, for example, can only take up to 0.100 mol H^+ or OH^-. After that, if, say, 0.200 mol H^+ is added, then 0.100 mol H^+ is unreacted and the pH of the resulting solution is reflected in the concentration of the H^+ in the system that now has 0.100 mol H^+.

2. *Exercises*

a. A buffer is prepared by mixing 3.00 g of benzoic acid ($HC_7H_5O_2$) and 4.50 g of sodium benzoate ($NaC_7H_5O_2$) in 2.00 L of water. The K_a for benzoic acid is 6.6×10^{-5}.

(1) Calculate the pH of the buffer.

Solution: We will use the equilibrium system of the weak acid ionization.

$$HC_7H_5O_2 \text{ (aq)} \rightleftharpoons H^+ \text{ (aq)} + C_7H_5O_2^- \text{ (aq)}$$

The equation to determine H^+ is

$$[H^+] = K_a \times \frac{\text{mol } HC_7H_5O_2}{\text{mol } C_7H_5O_2^-}$$

We now find the number of moles of weak acid and conjugate base.

$$\text{mol } HC_7H_5O_2 = 3.00 \text{ g} \times \frac{1 \text{ mol } HC_7H_5O_2}{122 \text{ g } HC_7H_5O_2} = 0.0246$$

$$\text{mol } NaC_7H_5O_2 = 4.50 \text{ g} \times \frac{1 \text{ mol } NaC_7H_5O_2}{144 \text{ g } NaC_7H_5O_2} = 0.0313$$

Calculating $[H^+]$, we get

$$[H^+] = 6.6 \times 10^{-5} \times \frac{0.0246}{0.0313} = 5.2 \times 10^{-5} \text{ M}$$

The pH of the buffer is 4.28.

(2) Calculate the pH of the buffer when 10.0 mL of 0.100 M HCl is added.

Solution: We first calculate the number of moles of H^+ added.

$$0.0100 \text{ L} \times \frac{0.100 \text{ mol HCl}}{1 \text{ L}} \times \frac{1 \text{ mol } H^+}{1 \text{ mol HCl}} = 1.00 \times 10^{-3} \text{ mol } H^+$$

Since H^+ is added, we *increase* the number of moles of weak acid by 0.00100 mol.

$$\text{mol } HC_7H_5O_2 = 0.0246 + 0.00100 = 0.0256$$

The addition of H^+ *decreases* the number of moles of conjugate base by 0.00100 mol.

$$\text{mol } C_7H_5O_2^- = 0.0313 - 0.00100 = 0.0303$$

We can now calculate the pH of the buffer after the addition of HCl.

$$[H^+] = 6.6 \times 10^{-5} \times \frac{0.0256}{0.0303} = 5.6 \times 10^{-5}$$

The pH of the buffer after the addition of HCl is 4.25. Recall that before HCl was added, the pH was 4.28.

(3) Calculate the pH of the buffer when 0.500 g of NaOH is added to the buffer.

Solution: We calculate the moles of NaOH added, which is equal to the moles of OH^- added.

$$0.500 \text{ g NaOH} \times \frac{1 \text{ mol NaOH}}{40.0 \text{ g NaOH}} \times \frac{1 \text{ mol } OH^-}{1 \text{ mol NaOH}} = 0.0125 \text{ mol } OH^-$$

Remembering that the addition of OH^- *increases* the number of moles of conjugate base and *decreases* the number of moles of weak acid, we have

$$[H^+] = 6.6 \times 10^{-5} \times \frac{0.0246 - 0.0125}{0.0313 + 0.0125} = 1.8 \times 10^{-5}$$

The pH of the buffer after the addition of NaOH is 4.74.

(4) Calculate the pH of the buffer when 50.00 mL of 0.200 M HCl are added.
Solution: Again, we calculate the number of moles of strong acid (also equal to the number of moles of H^+) added.

$$0.05000 \text{ L} \times \frac{2.000 \text{ mol HCl}}{1 \text{ L}} \times \frac{1 \text{ mol } H^+}{1 \text{ mol HCl}} = 0.1000 \text{ mol } H^+$$

We see that we do not have enough weak base $C_7H_5O_2^-$ to react with H^+. Since we have only 0.0313 mol of weak base, we will have 0.1000 - 0.0313 = 0.0687 mol of unreacted H^+. The buffer has been destroyed. The concentration of H^+ in the resulting solution is

$$[H^+] = \frac{0.0687 \text{ mol}}{2.000 \text{ L} + 0.0500 \text{ L}} = 0.0335 \text{ M}$$

The pH of the solution is 1.47. Note the precipitous drop from a pH of 4.28.

b. A buffer is prepared by mixing 500.0 mL of 1.250 M ammonia and 25.00 g of ammonium chloride.

(1) Calculate the pH of the buffer. **(E3)**

(2) Calculate the pH of the buffer after 25.0 mL of 0.500 M HCl are added to the buffer. **(E4)**

(3) Calculate the pH of the buffer after 10.0 g of $Ba(OH)_2$ are added to the buffer. **(E5)**

(4) Calculate the pH of the buffer after 0.2500 L of 1.500 M $Ba(OH)_2$ are added to the buffer. **(E6)**

II. Acid-Base Indicators
A. Nature of an indicator
1. What does it do? _____

_____

2. What is it derived from? _____

_____

C. Color of an indicator
1. When $[HIn]/[In^-] \geq 10$:

    a. The principal species is _____ .

    b. The color observed is that of _____ .

2. When $[HIn]/[In^-] \leq 0.10$:

    a. The principal species is _____ .

    b. The color observed is that of _____ .

3. When $[HIn] \approx [In^-]$:

    The color observed is that of _____ .

4. Color is dependent on

    a. pH of the solution

    b. K_a of the indicator

III. Acid-Base Titrations
A. Strong acid-strong base titration
1. Give an example and write an equation for the reaction that takes place.

_____

_____

2. Draw the titration curve.

3. pH at the equivalence point: _____

4. Species present in solution at the equivalence point: _____

5. Types of indicators that can be used for this titration: _____

6. Calculating the pH of the solution at different points in the titration

In these titrations we will concentrate on three points:

a. Starting point — when no titrant has been added

b. Half way to the equivalence point

c. At the equivalence point

7. *Exercises*

a. If 25.00 mL of 0.200 M $HClO_4$ is titrated with 0.100 M KOH, find the pH of the solution at the following points.

(1) When zero titrant has been added (starting point)

Solution: The titrant is the species that is being added. It is therefore the reagent whose volume is not initially given, in this case KOH. Before KOH is added, we have 25.00 mL of a strong acid (0.200 M $HClO_4$). Recall from Chapter 13 that the [H^+] of a strong acid is simply its molarity. Thus

$$[H^+] = 0.200\,M \quad \text{and} \quad pH = 0.699$$

(2) Half way to the equivalence point

Solution: In this case, it is a good idea to first treat acid and base separately. Determine the volume, molarity and number of moles of one reactant, and the volume, molarity and number of moles of the titrant needed to react with half the number of moles of the first.

H^+ (from $HClO_4$):   Volume = 0.02500 L
                                Molarity = 0.200 M
                                moles = $0.02500 \times 0.200 = 0.00500$

OH^- (from KOH):   Volume = _____
                                Molarity = 0.100 M
                                moles = _____

To determine how many moles of OH^- have to be added to reach half-neutralization, we take half the moles of H^+. In this case

$$\text{moles } OH^- = 0.00500 \div 2 = 0.00250$$

We can then determine the volume of KOH added.

$$\text{volume} = \frac{0.00250\,\text{moles}}{0.100\,\dfrac{\text{moles}}{L}} = 0.0250\,L$$

Now we put the acid and base together. The reaction is

$$H^+\,(aq) + OH^-\,(aq) \;\rightarrow\; H_2O$$

Reacting 0.00500 moles of H^+ with 0.00250 moles of OH^- means that all the OH^- gets used up and the number of moles of H^+ left over is

$$0.00500 - 0.00250 = 0.00250 \text{ moles}$$

The total volume of the mixture is

$$0.0250\,L + 0.0250\,L = 0.0500\,L$$

Thus $[H^+]$ in the solution after KOH has been added half-way to the equivalence point is

$$[H^+] = \frac{0.00250 \text{ moles}}{0.0500\,L} = 0.0500\,M$$

The pH of the solution is 1.301.

(3) At the equivalence point
   Solution: No calculations are necessary. The pH at the equivalence point for a strong acid - strong base titration is always 7.

b. Forty mL of 0.250 M HNO_3 are titrated with 0.200 M $Ba(OH)_2$. Determine the pH of the solution

   (1) at the starting point (zero titrant).

   (2) half way to the equivalence point.

   (3) at the equivalence point. **(E7)**

B. Strong acid-weak base titration

  1. Give an example and write an equation for the reaction that takes place.

_____

_____

  2. Draw the titration curve.

  3. pH at the equivalence point: _____

  4. Species present in solution at the equivalence point: _____

  5. Types of indicators that can be used for this titration: _____

  6. Calculating the pH

   a. When no titrant has been added

   The titrant is almost always the strong acid. Thus, determining the pH before addition of the strong acid is simply a matter of determining the pH of a solution of weak base. Review Chapter 13 to see how that is done.

   b. At half-neutralization

   At half-neutralization, half of the weak base (B^-) has been converted to its conjugate acid (HB) by the reaction

$$B^- \text{ (aq)} + H^+ \text{ (aq)} \rightleftharpoons HB \text{ (aq)}$$

   Thus $[B^-] = [HB]$ and substituting into the equilibrium expression

$$K_b = \frac{[OH^-][HB]}{[B^-]}$$

   for the weak base we get

$$K_b = \frac{[OH^-][B^-]}{[B^-]} = [OH^-]$$

   So at half-neutralization $[OH^-] = K_b$ and the pH can be easily calculated from that information.

   c. At neutralization

   At this point, all the weak base (B^-) has been converted to HB. So determining the pH of the resulting solution is similar to the method discussed in Chapter 13 for the calculation of the pH of a weak acid.

Remember: The molarity of the weak acid formed is determined by taking the number of moles of weak base (which is equal to the number of moles of conjugate acid formed) and dividing by the total volume of the solution (volume of weak base + volume of strong acid titrant).

7. *Exercises*

    a. Consider the titration of sodium cyanide, NaCN, with HCl. Twenty-five mL of 0.200 M NaCN are titrated with 0.250 M HCl.

       (1) Calculate the pH when no titrant has been added.

         Solution: Before titration, the weak base CN^- ionizes in water according to the equation

$$CN^- \text{ (aq)} + H_2O \rightleftharpoons OH^- \text{ (aq)} + HCN \text{ (aq)}$$

The equilibrium constant expression is

$$K_b = \frac{[OH^-][HCN]}{[CN^-]}$$

We construct a table that looks like this:

	CN^-	+	H_2O	$\rightleftharpoons$	OH^-	+	HCN
$[\ ]_o$	0.200				0		0
$\Delta[\ ]$	$-x$				$+x$		$+x$
$[\ ]_{eq}$	$0.200-x$				x		x

Substituting into the K_b expression, we get

$$K_b = \frac{(x)(x)}{0.200 - x}$$

Making our usual assumptions and checking them, we find x:

$$1.7 \times 10^{-5} = \frac{x^2}{0.200}$$

$$x = [OH^-] = 0.0018 \, M$$

$$pOH = 2.73 \quad \text{and} \quad pH = 11.27$$

       (2) At half-neutralization

         Solution: At half-neutralization, $K_b = [OH^-]$. Thus

$$[OH^-] = 1.7 \times 10^{-5}$$

$$pOH = 4.77 \quad \text{and} \quad pH = 9.23$$

(3) At the equivalence point

Solution: Here we have to determine the volume of HCl required to reach the equivalence point. To do that, recall that to reach the equivalence point the same number of moles of acid must be added to the moles of weak base in the solution. Hence

$$\text{mol } H^+ = \text{mol } CN^- = 0.0250\,L \times 0.200\,\frac{\text{moles}}{L} = 0.00500\text{ moles}$$

Since $[H^+] = [HCl]$, then the volume of HCl can be calculated.

$$V_{HCl} = 0.00500\text{ mol HCl} \times \frac{1\,L\text{ HCl}}{0.250\text{ mol HCl}} = 0.0200\,L$$

We now find the molarity of the weak acid (HCN) formed. Recall again that the number of moles of HCN formed is equal to both the number of moles of weak base (CN^-) and the number of moles of strong acid (H^+). The volume of the resulting solution is the sum of the volumes of weak base and strong acid.

$$V_{HCN} = 0.0250\,L\text{ (from } CN^-) + 0.0200\,L\text{ (from HCl)} = 0.0450\,L$$

We now calculate molarity

$$M = \frac{\text{moles HCN}}{V_{HCN}} = \frac{0.00500\text{ mol}}{0.0450\,L} = 0.111$$

Knowing the molarity of the weak acid, we again construct a table and proceed to calculate the pH of the weak acid.

	HCN	$\rightleftharpoons$	H^+	+	CN^-
$[\ ]_0$	0.111		0		0
$\Delta[\ ]$	$-x$		$+x$		$+x$
$[\ ]_{eq}$	$0.111 - x$		x		x

We substitute the values in the third row into the equilibrium expression

$$K_a = \frac{[H^+][CN^-]}{[HCN]}$$

and get

$$5.8 \times 10^{-10} = \frac{(x)(x)}{0.111 - x}$$

Making the usual assumptions and checking them, we have

$$5.8 \times 10^{-10} = \frac{x^2}{0.111}$$

Solving for x, we obtain

$$x = 8.0 \times 10^{-6} = [H^+]$$

The pH is 5.10.

b. A solution of ammonia is titrated with hydrobromic acid. If 30.00 mL of 0.100 M NH_3 are titrated with 0.200 M HBr, calculate the pH

(1) at the starting point (no titrant).

(2) half way to the equivalence point.

(3) at the equivalence point. **(E8)**

C. Weak acid-strong base titration

1. Give an example and write an equation for the reaction that takes place.

_____

_____

2. Draw the titration curve.

3. pH at the equivalence point: _____

4. Species present in solution at the equivalence point: _____

5. Types of indicators that can be used for this titration: _____

6. Calculating the pH of the solution at different points in the titration
   a. When no titrant has been added
   This time the titrant is the strong base. Determine the pH of the weak acid. Review Chapter 13.
   b. At half-neutralization
   Using the same argument we made in the strong acid - weak base titration, we see that $K_a = [H^+]$ at half-neutralization.
   c. At the equivalence point
   This time a weak base is formed. $[OH^-]$ is calculated using the ionization constant equation for K_b. Again, remember that you need the total volume of the solution (after an equivalent amount of titrant has been added) in order to calculate the molarity of the weak base.

7. Twenty-five mL of 0.400 M propionic acid ($HC_3H_5O_2$) solution are titrated with 0.300 M NaOH. Calculate the pH. (K_a for propionic acid = 1.4×10^{-5}.)
   a. at the starting point (no titrant).

   b. half way to the equivalence point.

   c. at the equivalence point. **(E9)**

SELF-TEST
A. Multiple Choice:

1. The K_a of benzoic acid is 1×10^{-5}. Which of the following would you choose as a pH = 4.0 buffer?
   a. A 1.0 M solution of benzoic acid
   b. A 1.0 L solution containing 1.0 mole of benzoic acid and 1.0 mole of sodium benzoate
   c. A 1.0 L solution containing 1.0 mole of benzoic acid and 0.10 mole of sodium benzoate
   d. A 1.0 L solution containing 0.10 mole of benzoic acid and 1.0 mole of sodium benzoate

2. Which of the following indicators is most suitable for the titration of 0.1 M ammonia with 0.1 M HCl?

a. cresol red	pH range	0.2 – 1.2
b. methyl red	pH range	4.2 – 6.2
c. neutral red	pH range	7.0 – 8.2
d. thymol blue	pH range	8.0 – 9.6
e. alizarin yellow	pH range	10.1 – 12.6

3. A buffered solution contains 0.100 mol sodium acetate and 0.100 mol acetic acid. Its pH is 4.7. One gram of NaOH (0.025 mol) is added to the buffer. What is the pH of the solution after the NaOH has been dissolved?
   a. less than 4   **b.** near 4.7   **c.** 7.0   **d.** 12.4

4. Which of the following relationships applies to a solution made by adding 0.150 mol of weak base, B^-, to water to make one liter of solution?
   a. $[B^-] = [OH^-]$ **b.** $[B^-] = [HB]$
   c. $[B^-] + [OH^-] = 0.15$ M   **d.** $[HB] + [OH^-] = 0.15$ M

5. One liter solutions with the following reagents are prepared. Which ones are buffers?

    (1) 0.05 mol HCl and 0.1 mol NH_3
    (2) 0.1 mol HCl and 0.05 mol NH_3
    (3) 0.1 mol NH_4Cl and 0.05 mol NH_3
    (4) 0.1 mol NaOH and 0.05 mol NH_4Cl
    (5) 0.05 mol NaOH and 0.1 mol NH_4Cl

a. (3)   **b.** (1),(2),(3)   **c.** (2),(3),(4)
d. (1),(3),(5)   **e.** (1),(2),(3),(4),(5)

6. The indicator phenolphthalein is red at a pH greater than 9. Which of the following solutions would turn red when phenolphthalein is added?

    (1) 1×10^{-10} M solution of HCl
    (2) 1×10^{-2} M solution of NaOH
    (3) equal volumes of 1×10^{-2} M solutions of NaOH and NaF
    (4) equal volumes of 1×10^{-2} M solutions of NaOH and HNO_3

a. (2)   **b.** (1),(2)   **c.** (2),(3)
d. (2),(4)   **e.** (1),(2),(3)   **f.** (1),(2),(4)

7. The indicator methyl red has a K_a of about 1×10^{-5}. Its "acid" form is red, while its "basic" form is yellow. What is the color of a solution with a pH of 5 to which a few drops of methyl red have been added?

a. red   **b.** yellow   **c.** green   **d.** orange   **e.** colorless

8. Which of the following equations best describes what takes place *just after the equivalence point is reached* in the titration of the weak base ClO^- (aq) with hydrobromic acid?

a. H^+ (aq) $+ ClO^-$ (aq) $\rightarrow$ HClO (aq)
b. HClO (aq) $\rightleftharpoons H^+$ (aq) $+ ClO^-$ (aq)
c. ClO^- (aq) $+ H_2O \rightleftharpoons$ HClO (aq) $+ OH^-$ (aq)
d. HClO (aq) $+ OH^-$ (aq) $\rightarrow H_2O + ClO^-$ (aq)
e. ClO^- (aq) $+ Br^-$ (aq) $\rightarrow ClOBr^{2-}$ (aq)

9. Which of the following is the net ionic equation for the reaction that occurs when small amounts of HCl (aq) are added to the $HNO_2/NaNO_2$ system?

a. H^+ (aq) $+ H_2O \rightarrow H_3O^+$
b. H^+ (aq) $+ NO_2^-$ (aq) $\rightarrow HNO_2$ (aq)
c. HNO_2 (aq) $\rightarrow H^+$ (aq) $+ NO_2^-$ (aq)
d. H^+ (aq) $+ HNO_2$ (aq) $\rightarrow H_2NO_2^+$ (aq)
e. HCl (aq) $+ NaNO_2$ (aq) $\rightarrow HNO_2$ (aq) $+$ NaCl (aq)

10. Which of the following statements is true about the titration of a weak monoprotic base with a strong acid such as HNO_3? At the equivalence point
   a. the pH is 7.
   b. the number of moles of weak base is greater than the number of moles of strong acid added.
   c. phenolphthalein ($K_a \approx 1 \times 10^{-10}$) turns pink.
   d. the resulting solution is a buffer.
   e. the resulting solution is acidic.

B. Less than, Equal to, Greater than

Consider 2 beakers. Beaker A has 50.0 mL of 0.10 M HCl. Beaker B has 50.0 mL of 0.10 M $HC_2H_3O_2$ ($K_a = 1.8 \times 10^{-5}$). Both are to be titrated with a solution of 0.10 M NaOH. Answer questions 1 – 4 below, using **LT** (for *less than*), **GT** (for *greater than*), **EQ** (for *equal to*), or **MI** (for *more information required*) in the blanks provided.

_____ 1. At the equivalence point, the volume of NaOH used to titrate HCl in Beaker A is ___(1)___ the volume of NaOH used to titrate $HC_2H_3O_2$ in Beaker B.

_____ 2. When each solution has reached its equivalence point, the pH of the solution in Beaker A is ___(2)___ the pH of solution in Beaker B.

_____ 3. Before titration starts (at zero time), the pH of the solution in Beaker A is ___(3)___ the pH of the solution in Beaker B.

_____ 4. At half-neutralization (halfway to the equivalence point), the pH of the solution in Beaker A is ___(4)___ the pH of the solution in Beaker B.

_____ 5. The effective range for the indicator used in the titration of Beaker B ___(5)___ the effective range for the indicator used in the titration of Beaker A.

Consider the following titrations. What would the pH be at the equivalence point with respect to 7?

_____ 6. HCl with NaF

_____ 7. $HCHO_2$ with KOH

_____ 8. HCl with KOH

_____ 9. $HCHO_2$ with NaF

C. True or False:

Consider the titration of a weak acid with sodium hydroxide.

_____ **1.** The pH is 7 at the equivalence point.

_____ **2.** $[H^+]$ is less than $[OH^-]$ after the equivalence point is reached.

_____ **3.** As base is added the pH increases slowly then a large increase occurs.

_____ **4.** As base is added the pH increases until the half equivalence point is reached, then the pH remains constant as more base is added.

_____ **5.** At the half equivalence point the $[H^+] = K_a$.

_____ **6.** Before the half equivalence point the concentration of the weak acid is less than the concentration of its conjugate base.

D. Problems:

Consider ammonia, NH_3.

1. What is the pH of a buffer solution prepared by adding 15.0 g of ammonium chloride to 5.00 L of 0.200 M ammonia?

2. What is the pH of the buffer after 90.0 mL of 2.00 M HCl are added?

3. What is the pH of the buffer after 5.00 g of NaOH are added?

4. What is the pH of the buffer after 15.00 g of NaOH are added?

5. How many grams of ammonium chloride are needed for 5.00 L of 0.200 M ammonia to make a buffer with a pH of 10.50?

6. If 50.0 mL of 0.200 M solution of ammonia are titrated with 0.100 M solution of HNO_3, what volume of HNO_3 is required to reach the equivalence point? What is the pH at that point?

7. If the titration in (6) stops half way to the equivalence point, what is the pH of the solution at that point?

ANSWERS
Exercises:
(E1) 6.36
(E3) 9.38
(E5) 9.58
(E7) (1) 0.602   **(2)** 1.02   **(3)** 7.00
(E9) a. 2.63   **b.** 4.85   **c.** 9.04

(E2) 32 g
(E4) 9.36
(E6) 13.58
(E8) (1) 11.13   **(2)** 9.26   **(3)** 5.21

Self-Test
A. Multiple Choice:
1. c 2. b 3. b 4. c 5. d
6. c 7. d 8. b 9. b 10. e

B. Less than, Equal to, Greater than:
1. EQ 2. LT 3. LT 4. LT 5. GT
6. LT 7. GT 8. EQ 9. MI

C. True or False:
1. F 2. T 3. T 4. F 5. T
6. F

D. Problems:
1. 9.80 2. 9.50 3. 10.11 4. 12.28
5. 3.0 g 6. 0.100 L; 5.21 7. 9.26

15
Complex Ions

I. **Composition of Complex Ions**
   A. Nomenclature
      1. Coordinate covalent bond

      _____

      2. Complex ion

      _____

      3. Central metal atom

      _____

      4. Ligand

      _____

      5. Coordination number

      _____

   B. Complex ions in water solution
      1. Lewis acid

         a. Definition: _____

            _____

         b. Example: _____
      2. Lewis base

         a. Definition: _____

            _____

     b. Example: _____

C. Charges of complexes

    1. Charge of the complex ion

       a. The "formula" you plug into is

          charge complex ion = metal oxidation number + total ligand charge

       b. To get the total ligand charge, you multiply each subscript by the charge of its ligand and add them all up.

    2. Determining the oxidation number of the central metal atom.

       a. You can use the same general "formula" given above. This time you solve for the oxidation number of the central metal atom.

       b. *Exercises*

         (1) In the complex ion $[Co(NH_3)_4Cl_2]^+$, what is the oxidation number of cobalt?

           Solution: The charge of the complex ion is +1. It is the superscript at the end of the complex ion. (The plus sign by itself means +1). If there is no superscript, then the charge of the complex ion is 0. For this complex ion, the first ligand NH_3 is a molecule, so it has 0 charge. The second ligand Cl has a charge of -1. Plugging into our "formula", where x is the oxidation number of Co, we get

$$+1 = x + 4(0) + 2(-1)$$
$$x = 3$$

         The oxidation number of cobalt in this ion is +3.

        (2) Determine the oxidation number of the central metal atom in the following complex species:

         (a) $Zn(H_2O)_2(OH)Cl$

         (b) $[PtCl_4]^{2-}$ **(E1)**

    3. The formulas of coordination compounds

       a. Complex ions most often do not exist by themselves. They are either attached to anions (if the complex ion is a cation) or to cations (if the complex ion is an anion).

       b. The coordination compound is neutral.

c. To write the formula of a coordination compound, follow these steps:
  (1) Choose a simple anion or cation that has the same charge as the complex ion.
  (2) Write the formula of the simple ionic compound using the simple anion or cation that you picked in Step (1).
  (3) Write the formula for the coordination compound by copying the formula for the simple ionic compound and substituting the complex ion for the simple ion you picked in Step (1). Remember to enclose the complex ion in brackets.

d. *Exercises*
  (1) Write the formula for the coordination compound that is formed by barium and $Cr(CN)_6^{3-}$.
      Solution:
      (a) Since the charge on the complex anion $Cr(CN)_6^{3-}$ is –3, an analogous simple anion would be N^{3-}.
      (b) The simple ionic compound formed by Ba^{2+} and N^{3-} would be Ba_3N_2.
      (c) The coordination compound is $Ba_3[Cr(CN)_6]_2$.
          Remember: The charges of the ions are not included as superscripts, and the subscript for the entire complex ion is written outside the brackets.
  (2) Write the formula for the coordination compound that is formed by $Zn(H_2O)_4^{2+}$ and phosphate ion. **(E2)**

      After sufficient practice, you will no longer consciously have to follow the steps to write coordination compound formulas.
D. Determining the coordination number
  1. Ligands
    a. Theoretically, any molecule or anion with an unshared pair of electrons that can form a coordinate covalent bond with a metal ion can be a ligand.
    b. Practically, a ligand usually contains an atom of the following elements: C (CN^-), N (NH_3), O (OH^- or H_2O), S (H_2S), F (F^-), Cl (Cl^-), Br (Br^-), I (I^-).
    c. Kinds of ligands
      (1) Monodentate
          Definition: _____
      (2) Bidentate
          Definition: _____
      (3) Polydentate
          Definition: _____
          You should memorize the following polydentate ligands:
          (a) Ethylenediamine (en) — bidentate
          (b) Oxalate (ox) — bidentate
          (c) EDTA — polydentate

2. Coordination number
   It is the number of bonds formed by the central metal atom.
3. *Exercises*
   a. What is the coordination number of the central metal atom in $Co(en)_2ox^+$?
   Solution: Since both en and ox are bidentate, each ligand contributes two bonds. There are three ligands (two en and one ox); thus there are six bonds in all and the metal has coordination number 6.
   b. Give the coordination number of the central metal atom in
   (1) $Al(OH)_2(ox)(NH_3)_2^-$

   (2) $Au(H_2O)(OH)F_2^{2-}$ **(E3)**

II. Geometry of Complex Ions
A. The geometry of complex ions is dependent on the coordination number.
B. Coordination number = 2
   1. The geometry is linear.
   2. No geometric isomers are possible.
C. Coordination number = 4
   1. The geometry can be tetrahedral.
      No geometric isomers are possible.
   2. The geometry can be square planar.
      Geometric isomerism is possible if the complex ions are of the type Ma_2b_2 or Ma_2bc, where M is the metal atom and a, b, and c are the ligands.
      a. Trans isomerism
         Example:

      b. Cis isomerism
         Example:

D. Coordination number = 6

1. The geometry is octahedral.
2. The isomerism can be cis or trans.
   It is difficult to see whether one structure is really an isomer or just a rotation of the same structure. It sometimes helps to redraw the figure after rotating it 90°.
3. Note that the two bonds of a bidentate ligand (like en) must always be cis to each other. It is impossible for the two bonds of the same molecule to occupy positions trans to each other.

E. *Exercises*

1. How many isomers are there in the complex $Co(H_2O)_2(NH_3)_2Cl_2^+$?
   Solution: We start by putting the water molecules trans to each other while the ammonia molecules and chloride ions are cis to each other.

Then we put the chloride ions trans to each other, while the ammonia and water molecules are cis to each other.

Now we put the ammonia molecules trans to each other, while the water molecules and chloride ions are cis to each other.

Next, we put all the ions and molecules cis to each other.

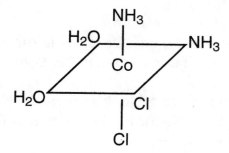

We end by putting all the ions and molecules trans to each other.

$$
\begin{array}{c}
H_2O \\
Cl\!-\!|\!-\!-\!-\!NH_3 \\
Co \\
NH_3\!-\!-\!-\!Cl \\
H_2O
\end{array}
$$

2.   How many isomers are there in the square planar ion $Cu(H_2O)_2(NH_3)_2^+$? **(E4)**

3.   How many isomers are there in the neutral complex $Ni(en)_2Cl_2$? **(E5)**

III. Electronic Structure and Bonding in Complex Ions
   A. Electron configuration of the central metal atom
   When transition metals form their ions, the s electrons are the first ones removed, and then the d electrons.
   B. Crystal field model
   1. The assumption is that the bonding between ligands and the metal ion is primarily electrostatic in nature.
   2. An electrostatic field is created around the d orbitals of the metal ion by the ligands. This field changes the relative energies of the different d orbitals.

3. Before interaction between ligands and the metal ion, all the d orbitals have the same energy. After interaction, the d orbitals are split into two groups with different energies.

   a. High-spin complexes

      (1) They contain the larger number of unpaired electrons.

      (2) Hund's rule is followed.

      (3) Δ_0 is small.

   b. Low-spin complexes

      (1) They contain the smaller number of unpaired electrons.

      (2) Electrons first pair up in lower energy orbitals.

      (3) Δ_0 is large.

4. The magnitude of the splitting energy Δ_0 determines the wavelength of light absorbed by a complex, and hence determines its color.

5. *Exercises*

   a. Using the crystal field model, derive the structure of the Ni^{3+} ion in the high-spin and low-spin octahedral complexes.

   Solution: We first determine the number of d electrons available. Nickel has 28 electrons; Ni^{3+} has 25 electrons and since 18 are in the argon core, there are 7 electrons left in the d orbitals.

   (1) High-spin complex

   We draw the 5d orbitals, putting the three low energy orbitals below the two high energy orbitals.

   ( )( )
   ( )( )( )

   We now distribute one electron in each of the five orbitals following Hund's rule. The model will now look like this:

   (↓)(↓)
   (↓)(↓)(↓)

   The remaining two electrons are used to fill the lower energy orbitals. You now have the structure of the high-spin complex.

   (↓)(↓)
   (↓ ↑)(↓ ↑)(↓)

   (2) Low-spin complexes

   Again, we draw the five orbitals as we did in part (1).

   ( )( )
   ( )( )( )

   This time, we disregard Hund's rule and put as many electrons as possible into the lower energy orbitals. Since there are three orbitals, and two electrons can fit into each orbital, we use up six of the seven electrons. The remaining electron goes into the high energy orbital. The structure of the low-spin complex looks like this:

$$(\downarrow\ )(\ )$$
$$(\downarrow\uparrow)(\downarrow\uparrow)(\downarrow\uparrow)$$

b. Using the crystal field model, derive the structure of the Co^{3+} ion in the high-spin octahedral complex, $[CoF_6]^{3-}$ and the low-spin octahedral complex $[Co(NH_3)_6]^{3+}$. **(E6)**

IV. Formation Constants of Complex Ions

A. Nature

1. The equilibrium constant for the formation of a complex ion or complex species is called a <u>formation constant</u> (or stability constant).
2. Symbol for the formation constant: _____

B. Writing equilibrium expressions

1. The equilibrium constant, K_f, is used for reactions at equilibrium where the complex ion is the product while the central atom and ligands are the reactants.
2. For the complex ion $[PtCl_2(NH_3)_2]^{2+}$, the equation for its formation is

$$Pt^{4+}\ (aq)\ +\ 2\,Cl^-\ (aq)\ +\ 2\,NH_3\ (aq)\ \rightleftharpoons\ Pt(NH_3)_2Cl_2^{2+}\ (aq)$$

3. The equilibrium expression for the reaction is

$$K_f\ =\ \frac{[Pt(NH_3)_2Cl_2^{2+}]}{[Pt^{4+}][Cl^-]^2\,[NH_3]^2}$$

4. *Exercise*

Write the equation for formation of the complex ion $[Co(CN)_2(NO_2)_2(H_2O)_2]^-$. Write the equilibrium expression for the equation. **(E7)**

C. Calculations using formation constants

These calculations are pretty straightforward. Just remember that complex ions must be made up of the metal ions and ligands.

1. *Example:* What is the ratio $[Cd(OH)_4^{2-}]/[Cd^{2+}]$ in a solution with a pH of 12.0?

   Solution: The equation for the formation of the complex ion is

   $$Cd^{2+}(aq) + 4OH^-(aq) \rightleftharpoons Cd(OH)_4^{2-}(aq)$$

   The equilibrium expression is

   $$K_f = 1.2 \times 10^9 = \frac{[Cd(OH)_4^{2-}]}{[Cd^{2+}][OH^-]^4}$$

   The rest is simply "plugging" in $[OH^-]$ which can be obtained from the pH. In this case, since the pH is 12.0. then $[OH^-] = 0.010$ M. The ratio then is

   $$\frac{[Cd(OH)_4^{2-}]}{[Cd^{2+}]} = (1.2 \times 10^9)(1.0 \times 10^{-2})^4 = 12$$

2. *Exercise:* At what pH is the $[Al(OH)_4^-]/[Al^{3+}]$ ratio 1.00×10^{20}? **(E8)**

SELF-TEST

A. Multiple Choice:

1. $[Au(NH_3)_2Cl_2]^+$ exists in two isomeric forms. This means that the complex structure must be

   a. linear   **b.** tetrahedral   **c.** square planar   **d.** octahedral

2. How many different geometric structures could be written for $[Rh(Br)_3(H_2O)_3]$?

   a. 1   **b.** 2   **c.** 3   **d.** 4   **e.** 5

3. The outer electron configuration for Cr^{3+} is

   a. $3d^5 4s^1$ **b.** $3d^4 4s^2$ **c.** $3d^3$ **d.** $3d^1 4s^2$

4. In the coordination compound $NH_4[Os(NH_3)_3Cl_3]$, what is the oxidation number of the central metal atom, osmium?

   a. +1   **b.** +2   **c.** +3   **d.** +4

5. Which of the following does not act as a ligand?
   a. $C_2O_4^{2-}$ **b.** NH_4^+ **c.** CO
   d. CN^- **e.** $NH_2-CH_2-CH_2-NH_2$

6. An octahedron has six corners. It is constructed from eight equilateral triangles. The length of a side is equal to s. Which of the following statements is true?
   a. From any corner, the distance to all other corners is s.
   b. From any corner, the distance to four other corners is s, while the distance to the fifth corner is greater than s.
   c. From any corner, the distance to three other corners is s, while the distance to the other two corners is greater than s.
   d. From any corner, the distance to two other corners is s, while the distance to the three other corners is greater than s.

7. In which pair of complex ions do the metals have the same oxidation number?
   a. $Pt(NH_3)_3Cl^+$; AlF_6^{3-} **b.** $Cu(en)_2(NH_3)_2^{2+}$; $Co(NH_3)_6^{3+}$

   c. $Cu(NH_3)_2^+$; $AuCl_4^-$ **d.** $Co(NH_3)_6^{3+}$; AlF_6^{3-}

8. Which of the following are bidentate ligands?
   (1) OH^- **(2)** H_2N-NH_2 **(3)** $H_2N-(CH_2)_2-NH_2$ **(4)** SCN^- **(5)** $C_2O_4^{2-}$

   a. (1),(2)   **b.** (2),(3)   **c.** (3),(5)   **d.** (2),(5)   **e.** (1),(4)

9. The number of molecules of hydrazine (NH_2-NH_2) and molecules of ammonia (NH_3) required to fill the six coordination positions in Cr^{3+} is
   a. 3 hydrazine molecules.
   b. 2 hydrazine molecules, 2 ammonia molecules.
   c. 2 hydrazine molecules, 4 ammonia molecules.
   d. 1 hydrazine molecule, 4 ammonia molecules.

10. The color of transition metal atoms in complexes can be explained by
   a. the crystal field model.
   b. the size of their ligands.
   c. the presence of geometric isomerism.
   d. all of the above.
   e. none of the above.

B. True or False:

_____ **1.** All metal ions exhibit only a single characteristic coordination number.

_____ **2.** A bidentate ligand has only one bond formed to two metal atoms.

_____ **3.** In geometric isomers, ligands that are cis to each other are closer together than those that are trans to each other.

_____ **4.** A square planar complex ion of the form Ma_2b_2, where M is the metal, and a and b are ligands, cannot have isomers.

_____ **5.** Ni^{3+} can have an electron distribution for both high spin and low spin complexes.

C. Fill in the blanks:

Consider $[Pd(en)_2(Cl)_2]^{2+}$.

_____ **1.** The coordination number for the complex ion is __(1)__ .

_____ **2.** Its geometry is __(2)__ .

_____ **3.** The oxidation number for Pd is __(3)__ .

_____ **4.** The formula for the phosphate salt of this ion is __(4)__ .

_____ **5.** Write the abbreviated electron configuration for the central metal ion. __(5)__

_____ **6.** If $[Pd(en)_2(Cl)_2]^{2+}$ is a high spin complex, how many unpaired electrons are there? __(6)__

D. Problems:

Chromium(III) ion forms a complex containing three water molecules and three chloride ions.

1. What is the charge of the complex species?

2. Using the crystal field model, give the electron distribution for the high spin complex and low-spin complex of Cr^{2+}.

3. How many isomers will the complex have?

4. Draw the structure of each isomer.

ANSWERS
Exercises:
(E1) +2; +2   **(E2)** $[Zn(H_2O)_4]_3(PO_4)_2$
(E3) 6; 4   **(E4)** 2
(E5) 2   **(E6)** high-spin complex   low-spin complex

high-spin complex	low-spin complex
(↓)(↓)	( )( )
(↓ ↑)(↓)(↓)	(↓ ↑)(↓ ↑)(↓ ↑)

(E7) $Co^{3+} (aq) + 2\,CN^- (aq) + 2\,NO_2^- (aq) + 2\,H_2O \rightleftharpoons [Co(CN)_2(NO_2)_2(H_2O)_2]^- (aq)$

$$K_f = \frac{[Co(CN)_2(NO_2)_2(H_2O)_2^-]}{[Co^{3+}]\,[CN^-]^2\,[NO_2^-]^2}$$

(E8) 10.8

Self-Test
A. Multiple Choice:

1. c	**2.** b	**3.** c	**4.** b	**5.** b
6. b	**7.** d	**8.** c	**9.** c	**10.** a

B. True or False:

1. F	**2.** F	**3.** T	**4.** F	**5.** T

C. Fill in the Blanks:

1. 6	**2.** octahedral	**3.** +4
4. $[Pd(en)_2(Cl)_2]_3(PO_4)_2$	**5.** $[Kr]\,4d^6$	**6.** 4

D. Problems:

1. 0

2. high-spin complex   low-spin complex   **3.** 2

high-spin complex:
(↑)( )
(↑)(↑)(↑)

low-spin complex:
( )( )
(↓ ↑)(↑)(↑)

4.

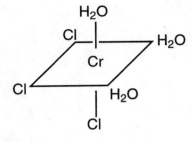

Precipitation Equilibria

I. Precipitation Equilibria

A. What is K_{sp}?

1. Definition: _____

_____

2. The equation showing the equilibrium between the solid solute and the dissolved ions in solution for a solid MX_2 is

$$MX_2 \text{ (s)} \;\rightleftharpoons\; M^{2+} \text{ (aq)} + 2X^- \text{ (aq)}$$

3. The equilibrium for the solubility equation written above is

$$K_{sp} = [M^{2+}][X^-]^2$$

4. The equilibrium constant K_{sp} follows the same rules as the equilibrium constant K for gaseous systems.
  a. It is constant for a given temperature.
  b. It is independent of the concentration of individual ions.
  c. It is the constant for the equation as written.
  Remember: When you are given the K_{sp}, the implied equilibrium equation (it is often not given) is that the solid is on the left side and the ions are on the right side, and the temperature at equilibrium is 25°C.

5. _Exercises_
  _a._ Write the K_{sp} expression for Al_2O_3.
  Solution: You must first write the ionization equation. Maybe later you will have enough practice so that you could eliminate this step. But for now it is _essential_ that you write the equation.

$$Al_2O_3 \text{ (s)} \;\rightleftharpoons\; 2Al^{3+} \text{ (aq)} + 3O^{2-} \text{ (aq)}$$

The K_{sp} expression is

$$K_{sp} = [Al^{3+}]^2 [O^{2-}]^3$$

Remember: In equilibrium expressions, solids and liquids are not included.

b. Write the K_{sp} expressions for nickel(II) chloride and potassium permanganate. **(E1)**

B. Uses of K_{sp}

1. Knowing the K_{sp} and the equilibrium concentration of one of the ions allows us to calculate the equilibrium concentrations of the other ions.

   Exercises

   a. A solution in equilibrium with barium phosphate solid has a barium ion concentration of 5×10^{-4} M. Barium phosphate has a K_{sp} of 6×10^{-39}. Calculate the phosphate ion concentration.

   Solution: First, you have to write the ionization equation for barium phosphate.

$$Ba_3(PO_4)_2 \text{ (s)} \rightleftharpoons 3\,Ba^{2+} \text{ (aq)} + 2\,PO_4^{3-} \text{ (aq)}$$

   To be able to do this, you must know how to write barium phosphate's formula correctly. Next, write the K_{sp} expression.

$$K_{sp} = 6 \times 10^{-39} = [Ba^{2+}]^3 [PO_4^{3-}]^2$$

   Since $[Ba^{2+}] = 5 \times 10^{-4}$

$$[PO_4^{3-}] = \left[\frac{6 \times 10^{-39}}{(5 \times 10^{-4})^3} \right]^{\frac{1}{2}} = 7 \times 10^{-15} \text{ M}$$

   b. Sulfuric acid (H_2SO_4) is added to a solution of lead nitrate. When equilibrium is established, $[SO_4^{2-}] = 2.0 \times 10^{-4}$. What is the equilibrium concentration of lead ion? **(E2)**

2. Knowing whether a precipitate will form
   a. When the original concentrations of the ions are given
      Exercises
      (1) A solution containing 0.020 M fluoride ions is added to a solution in which the original concentration of barium ions is 1.0×10^{-4} M. Will a precipitate form?

      Solution: First, we write the ionization equation assuming that a precipitate is likely to be formed between barium and fluoride ions.

$$BaF_2 \text{ (s)} \rightleftharpoons Ba^{2+} \text{ (aq)} + 2\,F^- \text{ (aq)}$$

Note that we write the ionization equation even if we are testing the likelihood of the *formation* of a precipitate.

Next, we write the ion product expression Q, which is the same as the K_{sp} expression except that original concentrations instead of equilibrium concentrations are used. This is analogous to the Q expression for K in Chapter 12.

$$Q = [Ba^{2+}]_o\, [F^-]_o^2$$

Substituting, we get a value for Q:

$$Q = (1.0 \times 10^{-4})\, (2.0 \times 10^{-2})^2 = 4.0 \times 10^{-8}$$

Compare Q with the K_{sp} for BaF_2 found in Table 16.1 of your text. Since $K_{sp} = 1.8 \times 10^{-7}$, $Q < K_{sp}$ and a precipitate will *not* form.

What concentration of fluoride ions is needed to get a saturated solution?

Solution: A saturated solution is one in which $Q = K_{sp}$. This means that we calculate $[F^-]$ in the K_{sp} expression by substituting values for the K_{sp} of BaF_2 and writing in the original Ba^{2+} concentration as the equilibrium concentration. Thus

$$1.8 \times 10^{-7} = (1.0 \times 10^{-4})\, [F^-]^2$$

Solving for $[F^-]$, we get

$$[F^-]^2 = \frac{1.8 \times 10^{-7}}{1.0 \times 10^{-4}}$$
$$[F^-] = 4.2 \times 10^{-2} \text{ M}$$

Thus, if the initial concentration of F^- is 4.2×10^{-2} M and the initial concentration of Ba^{2+} is 1.0×10^{-4} M, you get a saturated solution.

What concentration of fluoride ions gives a precipitate?

Solution: Any solution with the concentration of fluoride ions greater than 4.2×10^{-2} M gives a precipitate because then $Q > K_{sp}$.

(2) A solution originally contains 0.003 M magnesium ions. What concentration of hydroxide ions should be added to give an unsaturated solution, a saturated solution, a supersaturated solution? **(E3)**

b. When the volume and concentration of two solutions that are mixed are given

Exercises

(1) Will a precipitate form when 200.0 mL of 0.100 M aluminum nitrate is mixed with 300.0 mL of 0.200 M barium hydroxide? What is the precipitate?

Solution: First, you have to determine the possible precipitates. Then recall the solubility rules (Chapter 4) to decide the most likely candidate for equilibrium between a solid and its ions.

The possible precipitates are $Al(OH)_3$ and $Ba(NO_3)_2$. According to the solubility rules, the more likely precipitate is $Al(OH)_3$.

The question now is, are there enough aluminum and hydroxide ions in the solution for Q to exceed K_{sp}? We answer this question by first calculating the initial concentrations of Al^{3+} and OH^-. (We cannot use the given molarity for these ions, because a new solution with a different volume has been created by the mixture.) We start with the molarity of the aluminum ion.

$$0.200\,L \times \frac{0.100\,\text{mol } Al(NO_3)_3}{1\,L} \times \frac{1\,\text{mol } Al^{3+}}{1\,\text{mol } Al(NO_3)_3} = 0.0200\,\text{mol } Al^{3+}$$

$$M_{Al^{3+}} = \frac{0.0200\,\text{mol } Al^{3+}}{0.200\,L + 0.300\,L} = 0.0400$$

We do the same calculations to determine the molarity of the hydroxide ion.

$$0.300\,L \times \frac{0.200\,\text{mol } Ba(OH)_2}{1\,L} \times \frac{2\,\text{mol } OH^-}{1\,\text{mol } Ba(OH)_2} = 0.120\,\text{mol } OH^-$$

$$M_{OH^-} = \frac{0.120\,\text{mol } OH^-}{0.200\,L + 0.300\,L} = 0.240$$

Now, we write the ionization equation for the proposed precipitate.

$$Al(OH)_3 \text{ (s)} \rightleftharpoons Al^{3+} \text{ (aq)} + 3OH^- \text{ (aq)}$$

Note again that we write the *ionization* equation even if we are considering the possibility of *forming* a precipitate.

The Q equation for the ionization is

$$Q = [Al^{3+}] [OH^-]^3$$

We substitute the concentrations of the ions obtained above, and get

$$Q = (0.0400) (0.240)^3 = 5.53 \times 10^{-4}$$

The K_{sp} for $Al(OH)_3$ is 2×10^{-31}. Since Q is larger than K_{sp}, a precipitate will form.

(2) Will a precipitate form if 250.0 mL of 0.400 M ammonium phosphate are mixed with 450.0 mL of 0.125 M calcium chloride? If a precipitate forms, write the formula of the precipitate. **(E4)**

3. **Knowing which of two possible precipitates will form first**
   Exercises
   a. A solution contains 0.02 M of Mg^{2+} and 0.02 M of Sr^{2+}. Sufficient carbonate is added so that the carbonate ion concentration is 2×10^{-7} M. Will a precipitate form, and if so what is it?

   Solution: We have to calculate the ion product Q for both possibilities. For $MgCO_3$, it is

$$Q = [Mg^{2+}]_o [CO_3^{2-}]_o$$
$$= (0.02) (2 \times 10^{-7}) = 4 \times 10^{-9}$$

For $SrCO_3$, it is

$$Q = [Sr^{2+}]_o [CO_3^{2-}]_o$$
$$= (0.02) (2 \times 10^{-7}) = 4 \times 10^{-9}$$

We now compare the K_{sp}'s of each precipitate with their respective ion products. For $MgCO_3$, K_{sp} is 6.8×10^{-6} and Q is 4×10^{-9}, so $MgCO_3$ will not precipitate. For $SrCO_3$, K_{sp} is 5.6×10^{-10} and Q is 4×10^{-9}, so $SrCO_3$ will precipitate.

b. A solution contains 0.0040 M of Ba^{2+} and 0.0060 M of Pb^{2+}. What concentration of F^- will just precipitate one of the ions? Which one will precipitate first? **(E5)**

C. K_{sp} and water solubility

1. Solubility means the number of moles of solid that can be dissolved in one liter of solution.
2. Solubility is expressed in moles/L or M.
3. Solubility is not the same as K_{sp}.
4. Calculating K_{sp} given solubility

   Exercises

   a. The measured solubility of MgF_2 at 25°C is 2.6×10^{-4} mol/L. Calculate the K_{sp} for MgF_2 if MgF_2 dissolves according to the equation

   $$MgF_2 \text{ (s)} \rightleftharpoons Mg^{2+} \text{ (aq)} + 2\,F^- \text{ (aq)}$$

   Solution: Since we are told that 2.6×10^{-4} mol of MgF_2 dissolves and the equation shows that 1 mole of MgF_2 splits up into 1 mole of Mg^{2+} and 2 moles of F^-, the equilibrium concentrations are $[Mg^{2+}] = 2.6 \times 10^{-4}$ M and $[F^-] = 5.2 \times 10^{-4}$ M. K_{sp} can now be calculated.

   $$\begin{aligned} K_{sp} &= [Mg^{2+}]\,[F^-]^2 \\ &= (2.6 \times 10^{-4})\,(5.2 \times 10^{-4})^2 = 7.0 \times 10^{-11} \end{aligned}$$

   b. The solubility of barium oxalate (BaC_2O_4) in water at 25°C is 22 mg/L. What is the K_{sp} for barium oxalate? **(E6)**

5. Calculating solubility given K_{sp}

The more useful exercise concerning the relationship between molar solubilities and K_{sp} is determining how many grams of a solid one can dissolve in a given volume of solvent, given the K_{sp} of the solid.

Exercises

a. How many grams of silver chromate can you dissolve in water to make one liter of solution?

Solution: Silver chromate is written as Ag_2CrO_4 and dissolves in water according to the equation

$$Ag_2CrO_4\ (s) \rightleftharpoons 2\,Ag^+\ (aq) + CrO_4^{2-}\ (aq)$$

If the solubility for Ag_2CrO_4 is designated as s, then one forms 2s moles of Ag^+ and s moles of CrO_4^{2-}. The K_{sp} expression for Ag_2CrO_4 is

$$K_{sp} = [Ag^+]^2\,[CrO_4^{2-}]$$

Substituting the s designations in place of concentrations, we get

$$1 \times 10^{-12} = (2s)^2(s) = 4s^3$$

Solving for s, we obtain

$$s = \left(\frac{1 \times 10^{-12}}{4}\right)^{\frac{1}{3}} = 6 \times 10^{-5}\,M$$

This means that 6×10^{-5} moles of Ag_2CrO_4 can be dissolved in enough water to make one liter of solution. We can now change moles to grams.

$$\frac{6 \times 10^{-5}\,mol}{1\,L} \times \frac{331.7\;g\;Ag_2CrO_4}{1\;mol\;Ag_2CrO_4} = 2 \times 10^{-2}\,\frac{g\;Ag_2CrO_4}{L}$$

Many students find it unnatural to multiply s by the coefficient of the ion in the balanced ionization equation and also to raise s to that same power. They somehow believe that dealing with the coefficient once is enough, that is multiply or raise to the power but not do both. This is not the case. You must do *both* operations when working with s.

b. How many milligrams of silver phosphate can you dissolve in water to make 100.0 mL of solution? **(E7)**

D. Common ion effect

1. Qualitative aspect

   Because of Le Chatelier's principle, solids are less soluble in solvents with which they have a common ion. For example, silver chloride is less soluble in a solution of HCl than in pure water. Silver chloride and HCl have a common ion, Cl^-. The addition of this common ion drives the equilibrium

$$AgCl\,(s) \;\rightleftharpoons\; Ag^+\,(aq) + Cl^-\,(aq)$$

   to the left, resulting in the formation of more AgCl (thus making AgCl less soluble).

2. Quantitative considerations

   When the solubility of a salt in pure water was calculated, we used the symbol s to denote the molar concentration of ions that would result. Thus for AgCl,

$$AgCl\,(s) \;\rightleftharpoons\; Ag^+\,(aq) + Cl^-\,(aq)$$

   If s moles/L of AgCl would dissolve, then since neither Ag^+ nor Cl^- is present initially

$$[Ag^+] = [Cl^-] = s$$

   However, if you want to calculate the solubility of AgCl in 0.10 M NaCl, then $[Cl^-]$ would no longer be equal to s but to $[0.10 + s]$. This reflects the fact that there are two sources of chloride ion. One source is Cl^- from NaCl (0.10 M). The other is from the ionization of AgCl in solution (s).

   Exercises:

   a. What is the solubility (in grams/L) of lead(II) chromate in 0.30 M K_2CrO_4?

   Solution: Every mole of lead chromate, $PbCrO_4$, that dissolves forms 1 mol of Pb^{2+} and 1 mol of CrO_4^{2-}. Since there are no lead ions in the solvent

$$[Pb^{2+}] = s$$

   Chromate ion is the common ion. There are two sources of chromate ion. They are the chromate ions from the solute $PbCrO_4$ (s), and those from the solvent, 0.30 M K_2CrO_4, where $[CrO_4^{2-}]$ is 0.30 M. The total concentration of chromate ion is therefore

$$[CrO_4^{2-}] = [CrO_4^{2-}]_{PbCrO_4} + [CrO_4^{2-}]_{K_2CrO_4}$$
$$= s + 0.30$$

   Substituting into the K_{sp} expression

$$K_{sp} = [Pb^{2+}][CrO_4^{2-}] = 2 \times 10^{-14}$$

   we get

$$2 \times 10^{-14} = (s)(0.30 + s)$$

   This is a quadratic equation. Since the CrO_4^{2-} contribution from the ionization of the solute ($PbCrO_4$) is small ($\approx\ 1 \times 10^{-7}$ M), we assume that s is much smaller than 0.30, and hence

$$0.30 + s \approx 0.30$$

With that assumption we write

$$2 \times 10^{-14} = (s)(0.30)$$
$$s = 7 \times 10^{-14} \, M$$

This means that 7×10^{-14} moles of $PbCrO_4$ will dissolve in one liter of 0.30 M K_2CrO_4.

To answer the question posed in the problem, which asks for the solubility in g/L, we do a simple conversion.

$$\frac{7 \times 10^{-14} \text{ mol } PbCrO_4}{L} \times \frac{323 \text{ g}}{1 \text{ mol } PbCrO_4} = 2 \times 10^{-11} \text{ g/L}$$

b. How many grams of $Ca(OH)_2$ can be dissolved in 0.20 M $Ba(OH)_2$? How many grams can be dissolved in pure water? **(E8)**

E. Selective precipitation

1. Selective precipitation is often used to separate two cations (or anions). The separation is accomplished by adding an anion (or cation) that will precipitate one of the cations (or anions) but not the other.

2. Selecting an anion (or cation) that will separate two cations (or anions), one first looks at the K_{sp}'s of the prospective precipitates. To separate, for example, Ca^{2+} from Fe^{3+}, one could add the hydroxide ion. K_{sp} for $Ca(OH)_2$ is 4.0×10^{-6}, while that for $Fe(OH)_3$ is 3×10^{-39}. Which ion do you think will precipitate first?

_____.

3. Qualitative aspects:

   In order to figure out how much of the anion (or cation) to add,

   a. Calculate the anion (cation) concentration using the K_{sp} of the prospective precipitate

   b. Use Q to calculate the concentration of the anion (cation) that you must add.

4. *Exercises*

   a. A solution consists of 0.10 M $MgCl_2$ and 0.10 M $ZnCl_2$. To separate the two cations from each other, solid NaOH is added. The volume of the solution does not change.

(1) Which salt, $Mg(OH)_2$ or $Zn(OH)_2$, will precipitate first?

Solution: Look for the K_{sp} of both saltas in Table 16.1 of your text. Calculate $[OH^-]$ required to just precipitate both salts.

For $Mg(OH)_2$:

$$6 \times 10^{-12} = [Mg^{2+}][OH^-]^2$$

$$[OH^-]^2 = \frac{6 \times 10^{-12}}{0.10}$$

$$[OH^-] = 8 \times 10^{-6}$$

For $Zn(OH)_2$:

$$4 \times 10^{-17} = [Zn^{2+}][OH^-]^2$$

$$[OH^-]^2 = \frac{4 \times 10^{-17}}{0.10}$$

$$[OH^-] = 2 \times 10^{-8}$$

This calculation shows that a smaller amount of NaOH is required to precipitate Zn^{2+}. Thus $Zn(OH)_2$ will precipitate first.

(2) At what hydroxide concentration will $Zn(OH)_2$ first begin to precipitate?

Solution: As the calculations above indicate, $Zn(OH)_2$ will start to precipitate when the OH^- concentration is 2×10^{-8}.

b. A solution has 0.20 M NaOH and 0.15 M Na_3PO_4. To separate the two anions, solid calcium nitrate is added.

(1) Which calcium salt ($Ca_3(PO_4)_2$ or $Ca(OH)_2$) will precipitate first?

(2) At what calcium concentration will the salt chosen in (1) first begin to precipitate? **(E9)**

II. Dissolving Precipitates

   A. Strong Acid

   1. Precipitates that can be dissolved by strong acids

      a. metal hydroxides

         Reaction for dissolving $Fe(OH)_3$ in HCl:

_____

b. almost all metal carbonates
Reaction for dissolving $CaCO_3$ in HCl:

_____

c. many metal sulfides
Reaction for dissolving FeS in HCl:

_____

2. Determination of K for the solution process
Most equations written for dissolving precipitates are overall equations reflecting a two-step process:
— The metal hydroxide, carbonate, or sulfide breaks up into ions.
— The hydroxide, carbonate, or sulfide ions react with H^+ (contributed by the strong acid) forming H_2O, H_2CO_3 (eventually CO_2 and H_2O), and H_2S, respectively. The equilibrium constant for the reaction with H^+ is so large that it drives the overall reaction essentially to completion.

Exercises:

_a._ Write the overall equation and calculate K for the reaction in which $Ca(OH)_2$ is dissolved in HNO_3.

Solution: The individual steps in dissolving $Ca(OH)_2$ in HNO_3 are

$$Ca(OH)_2 \text{ (s)} \rightleftharpoons Ca^{2+} \text{ (aq)} + 2\,OH^- \text{ (aq)}$$

$$2\,OH^- \text{ (aq)} + 2\,H^+ \text{ (aq)} \rightleftharpoons 2\,H_2O$$

The overall equation, obtained by adding the two equations above, is

$$Ca(OH)_2 \text{ (s)} + 2\,H^+ \text{ (aq)} \rightleftharpoons 2\,H_2O + Ca^{2+} \text{ (aq)}$$

To calculate K for the overall reaction, consider the K's for the individual steps. The first step has K_{sp} for its equilibrium constant (K_1). The Ksp for $Ca(OH)_2$ is 4.0×10^{-6}. The second step has $1/K_w$ for its equilibrium constant (K_2). Since we doubled all the coefficients for the second equation, $K_2 = (1/K_w)^2$. Thus for the overall reaction

$$K = K_1 \times K_2 = K_{sp} \times \left(\frac{1}{K_w}\right)^2$$

$$= 4.0 \times 10^{-6} \times \left(\frac{1}{1.0 \times 10^{-14}}\right)^2 = 4.0 \times 10^{22}$$

b. Write the overall equation and calculate K for dissolving $Fe(OH)_3$ in strong acid. **(E10)**

3. Calculating the molar solubility for dissolving metal hydroxides in acid
   Exercises:
   a. Using the overall reaction and the value for K calculated above for dissolving $Ca(OH)_2$ in HNO_3, what is the molar solubility, s, for $Ca(OH)_2$ in 0.003 M HNO_3?
   Solution: The equilibrium expression for the overall equation

$$Ca(OH)_2 \text{ (s)} + 2H^+ \text{ (aq)} \rightleftharpoons 2H_2O + Ca^{2+} \text{ (aq)}$$

is

$$K = \frac{[Ca^{2+}]}{[H^+]^2}$$

Since for every mole of $Ca(OH)_2$ that dissolves one mole of Ca^{2+} is formed, $[Ca^{2+}] = s$. The concentration of the strong acid, HNO_3, is 0.003 M, so $[H^+]$ = 0.003 M. Substituting into the equilibrium expression above, we get

$$4.0 \times 10^{22} = \frac{s}{(0.003)^2}$$

and

$$s = 4 \times 10^{17} \text{ M}$$

This means that $Ca(OH)_2$ is completely soluble in a solution as dilute as 0.003 M HNO_3.
   b. What is the molar solubility of $Ni(OH)_2$ in 5×10^{-4} M $HClO_4$? K_{sp} for $Ni(OH)_2$ is 2.0×10^{-15}. **(E11)**

B. Complex Formation

    1. Reactions

        a. Using NaOH to dissolve the precipitate

            (1) The product for the overall equation is the hydroxide complex.

            (2) The reaction occurs in a two-step process:

                — The solid ionizes into its component ions.

                — The metal cation forms a complex with the hydroxide ion.

        b. Using NH_3 to dissolve the precipitate

            (1) The product for the overall equation is _____.

            (2) Describe the two-step process involved in dissolving the precipitate.

        c. K for a reaction of this type is

$$K = K_{sp} \times K_f$$

    2. Quantitative applications

    Calculations used to determine the molar solubility of precipitates by dissolving them in NH_3 are similar to the process described above.

    Exercises:

    a. Cadmium(II) oxalate, CdC_2O_4, has a K_{sp} of 1.5×10^{-8}. Cadmium ions readily combine with ammonia to form the complex ion $Cd(NH_3)_4^{2+}$ ($K_f = 2.8 \times 10^7$). Calculate the molar solubility of CdC_2O_4 in 0.010 M NH_3.

    Solution: We first write the reactions that make up the overall equation.

$$CdC_2O_4\,(s) \;\rightleftharpoons\; Cd^{2+}\,(aq) + C_2O_4^{2-}\,(aq)$$

$$Cd^{2+}\,(aq) + 4\,NH_3\,(aq) \;\rightleftharpoons\; Cd(NH_3)_4^{2+}\,(aq)$$

$$CdC_2O_4\,(s) + 4\,NH_3\,(aq) \;\rightleftharpoons\; Cd(NH_3)_4^{2+}\,(aq) + C_2O_4^{2-}\,(aq)$$

The equilibrium expression for the overall reaction is

$$K = \frac{[Cd(NH_3)_4^{2+}][C_2O_4^{2-}]}{[NH_3]^4}$$

The numerical value for K is

$$K = K_f \times K_{sp} = (2.8 \times 10^7)(1.5 \times 10^{-8}) = 0.42$$

For every mole of CdC_2O_4 that dissolves, one mole of $C_2O_4^{2-}$ and one mole of $Cd(NH_3)_4^{2+}$ are formed. Thus we can write

$$s = [C_2O_4^{2-}] = [Cd(NH_3)_4^{2+}]$$

Substituting into the equilibrium expression we get

$$0.42 = \frac{(s)\,(s)}{(0.010)^4}$$

$$s^2 = 4.2 \times 10^{-9}$$

$$s = 6.5 \times 10^{-5} \text{ M}$$

You can see that not all complex ion formation leads to increased solubility. The molar solubility of CdC_2O_4 in water is 1.2×10^{-4}.

b. Determine the molar solubility of CdC_2O_4 in 5.0 M NH_3 and compare with the solubility obtained above with 0.010 M NH_3. **(E12)**

c. How many grams of AgBr can be dissolved in 1.0 L of 2.5 M NH_3? AgBr in NH_3 forms $Ag(NH_3)_2^+$. Compare with the mass of AgBr that can be dissolved in 1.0 L of pure water. **(E13)**

SELF-TEST
A. Multiple Choice:

1. The solubility product constant K_{sp} of Ag_2CrO_4 is related to its solubility in water as $K_{sp} =$
   a. s^2 **b.** s^3 **c.** $4s^3$ **d.** s   **e.** none of those

2. The solubility product expression for $Zn(CN)_2$ is
   a. $[Zn^{2+}][CN_2^-]$ **b.** $[Zn^{2+}][CN^-]^2$ **c.** $[Zn^{2+}][2\ CN^-]$ **d.** $[Zn^{2+}][2\ CN^-]^2$

3. In which of the solutions listed below is AgCl least soluble?
   a. pure water   **b.** 0.20 M $CaCl_2$ **c.** 0.20 M $AgNO_3$ **d.** 0.20 M NaCl

4. Which carbonate would be the first to precipitate from a solution of 0.1 M in metal cation as CO_3^{2-} is added?

   a. $BaCO_3$ $(K_{sp} = 2.6 \times 10^{-9})$ **b.** $CaCO_3$ $(K_{sp} = 4.9 \times 10^{-9})$

   c. $FeCO_3$ $(K_{sp} = 3.1 \times 10^{-11})$ **d.** $MgCO_3$ $(K_{sp} = 4 \times 10^{-6})$

5. Addition of an aqueous solution of KCN to an aqueous solution of $NiCl_2$ results in the formation of a solid which dissolves upon addition of more KCN (aq). Which of the following statements is the most probable explanation for these observations?

   a. Ni^{2+} forms $Ni(CN)_2$ (s) and then forms $Ni(CN)_4^{2-}$.

   b. KCl (s) is formed then reacts with additional CN^- to give ClCN (aq).

   c. $Ni(OH)_2$ (s) forms and then dissolves as $KNi(OH)_3$.

   d. $Ni(CN)_2$ (s) forms but goes back into solution because addition of KCN makes the resulting solution more dilute and hence unsaturated.

   e. none of these

B. True or False:

What happens when 50.0 mL of 0.050 M $BaCl_2$ is added to 50.0 mL of 0.10 M K_2SO_4?

_____ **1.** KCl will precipitate.

_____ **2.** The inital concentration of Ba^{2+}, $[Ba^{2+}]_o = 0.025$ M after the two solutions are mixed, but before reaction occurs.

_____ **3.** If $[Ba^{2+}]_o \times [SO_4^{2-}]_o > [Ba^{2+}]_{eq} \times [SO_4^{2-}]_{eq}$, then precipitation occurs.

_____ **4.** $[Cl^-]_{eq} = [K^+]_{eq} = 5.0 \times 10^{-2}$ M

Consider $PbCl_2$. When it dissolves, the following reaction takes place.

$$PbCl_2 \text{ (s)} \rightleftharpoons Pb^{2+} \text{ (aq)} + 2\,Cl^- \text{ (aq)} \qquad \Delta H° > 0$$

_____ **5.** $PbCl_2$ can be dissolved by a strong acid.

_____ **6.** $PbCl_2$ can be at least partially dissolved by a strong base if $Pb(OH)_4^{2-}$ exists.

_____ **7.** It is easier to dissolve $PbCl_2$ at higher temperatures.

_____ **8.** The molar solubility of $PbCl_2$ (s) in pure water can be expressed as $2s^3$.

C. Fill in the blanks:

Consider a saturated solution of $M(OH)_2$ where M^{2+} is a metal cation. Dissolving $M(OH)_2$ in water is an endothermic process. On the blanks provided, write **P** if a precipitate forms and **NP** if no precipitate forms when the saturated solution is subjected to:

_____ **1.** The addition of MCl_2 solution where M^{2+} is the same metal cation

_____ **2.** The addition of HCl

_____ **3.** An increase in temperature

_____ **4.** The addition of water

_____ **5.** The addition of ammonia to form $M(NH_3)_4^{2+}$

D. Problems:

Consider manganese sulfide, MnS, and tin(II) sulfide. Manganese sulfide has a K_{sp} of 5×10^{-14}. The solubility of tin(II) sulfide is 2×10^{-14} mol/L.

1. Calculate the K_{sp} of tin(II) sulfide.

2. How many milligrams of MnS can be dissolved in enough water to make 10.0 L of solution?

3. Calculate $[S^{2-}]$ if $[Mn^{2+}]$ is 2×10^{-5}.

4. Will a precipitate form if 0.500 mol of Mn^{2+} are added to one liter of a solution containing 0.0010 mol of S^{2-} ions?

5. In a solution that contains 1.00×10^{-4} mol of Mn^{2+} and 1.00×10^{-4} mol of Sn^{2+}, which will precipitate first when 1.00×10^{-10} mol S^{2-} is added?

6. Calculate the molar solubility of MnS in 0.3 M H^+ and 0.2 M H_2S. Assume that for the reaction

$$MnS\,(s)\ +\ 2\,H^+\,(aq)\ \rightleftharpoons\ Mn^{2+}\,(aq)\ +\ H_2S\,(aq)$$

K is 5×10^6.

Consider copper. The K_{sp} for CuCl is 1.7×10^{-7}. The K_f for $CuCl_2^-$ is 1×10^5.

7. Write the equation for the reaction between a solution of $CuNO_3$ and NaCl where a precipitate of CuCl is obtained.

8. At what concentration of NaCl will precipitation occur if NaCl is added to a 0.10 M solution of $CuNO_3$?

9. Write the equation for the formation of $CuCl_2^-$ from the reaction between CuCl and additional aqueous NaCl.

10. What is K for the reaction in (9)?

11. Compare the molar solubility of CuCl in pure water to that in aqueous 2.0 M NaCl where $CuCl_2^-$ is formed.

ANSWERS
Exercises:
(E1) $K_{sp} = [Ni^{2+}][Cl^-]^2$; $K_{sp} = [K^+][MnO_4^-]$ **(E2)** 9.0×10^{-5}

(E3) $[OH^-] < 4 \times 10^{-5}$ M gives an unsaturated solution   **(E4)** yes, $Ca_3(PO_4)_2$
     $[OH^-] = 4 \times 10^{-5}$ M gives a saturated solution
     $[OH^-] > 4 \times 10^{-5}$ M gives a supersaturated solution

(E5) 6.7×10^{-3} M; BaF_2 first   **(E6)** 9.6×10^{-9}

(E7) 2 mg   **(E8)** 0.0019 g/L; 0.74 g/L

(E9) $Ca_3(PO_4)_2$; 4×10^{-11} M

(E10) $Fe(OH)_3 (s) + 3H^+ (aq) \rightarrow Fe^{3+} (aq) + 3H_2O$; $K = 3 \times 10^3$

(E11) 5×10^6 M   **(E12)** 16 M

(E13) 1 g in NH_3; 0.1 mg (1×10^{-4} g in water)

Self-Test
A. Multiple Choice:
   1. c   **2.** b   **3.** b   **4.** c   **5.** a

B. True or False:
   1. F   **2.** T   **3.** T   **4.** F   **5.** F
   6. T   **7.** T   **8.** F

C. Fill in the Blanks:
   1. P   **2.** NP   **3.** NP   **4.** NP   **5.** NP

D. Problems:
   1. 4×10^{-28} **2.** 0.2 mg
   3. 3×10^{-9} **4.** yes
   5. SnS   **6.** 2×10^6 M
   7. $Cu^+ (aq) + Cl^- (aq) \rightleftharpoons CuCl (s)$ **8.** 1.7×10^{-6} M
   9. $CuCl (s) + Cl^- (aq) \rightleftharpoons CuCl_2^- (aq)$ **10.** 2×10^{-2}
   11. in pure water: 4.1×10^{-4} M
        in 2.0 M Cl^-: 4×10^{-2} M

Spontaneity of Reaction

I. **Spontaneous Processes**
   A. The energy factor
      1. Many spontaneous processes proceed with a decrease in energy.
      2. Berthelot and Thomsen suggested that all spontaneous reactions are exothermic.
         a. This is true for *almost* all chemical reactions occuring at 25°C and 1 atm.
         b. This does not hold for many phase changes. Consider ice melting. It is endothermic but spontaneous.
         c. Many reactions that are endothermic and nonspontaneous at room temperature become spontaneous at higher temperature but are still endothermic.
         Note: If all this seems foreign to you, it might be wise to go back and review Chapter 8. Reread the chapter in your text and redo the problems in this workbook. You must master enthalpy before you can go on to problems on spontaneity.
   B. The randomness factor
      1. Nature tends to move from a more ordered to a more random state.
      2. A random state is more probable than an ordered state.
      3. Exercises
         Choose the state that is more random, and thus more probable.
         a. A lawn before the leaves are raked or after they are raked
            Solution: The lawn before the leaves are raked is more random. There are many more ways that the leaves can be arranged before raking.
         b. A house being built or after it is built **(E1)**

    c. A deck of cards before or after a successful game of solitaire **(E2)**

    d. A football team in the huddle or after **(E3)**

II. Entropy, S
  A. The nature of entropy
    1. It is a characteristic property.
    2. It is a measure of disorder or randomness.
      a. Substances, like gases, with a high degree of randomness (molecules in gases are in constant motion) have high entropies.
      b. Substances, like crystals, with a low degree of randomness (molecules in crystals hardly move at all) have low entropies.
    3. Entropy is a state function.

$$\Delta S = S_{final} - S_{initial}$$

    4. Factors that influence the amount of entropy that a system has in a particular state

      a. _____

      _____

      b. _____

      _____

      c. _____

      _____

    5. Third law of thermodynamics

    _____

    6. An increase or decrease in temperature does not affect entropy much if the system stays in the same phase. There is however a large drop in entropy when a system in the liquid phase _____. There is a large increase in entropy when a system in the liquid phase _____.
  B. Standard molar entropies
    1. Unlike enthalpy, for which only its change can be measured, the absolute molar entropies of pure substances can be determined.

2. The symbol for standard molar entropies of pure substances at 25°C and 1 atm is S°. Note that there is no Δ in front of it. It is not a change in entropy.
3. The unit for S° is J/K. Remember that the unit for ΔH is kJ. This distinction should be kept in mind to avoid problems later.
4. The standard molar entropies (S°) of substances are always positive.
5. The ΔH_f° for elements in their native state at 25°C is zero. The entropy for elements in their native state is not zero and has to be looked up.
6. H^+ has an S° of zero.
7. Ions in solution have been assigned an S° given in J/K at 25°C and a concentration of 1 M.

C. ΔS° for reactions
1. In a reaction we can calculate the change in entropy, which is why we now have a Δ before the S°.
2. The relationship used to calculate ΔS° parallels that of ΔH°.

$$\Delta S^\circ = \sum S^\circ_{products} - \sum S^\circ_{reactants}$$

3. ΔS° and reactions involving gases
   a. A reaction that results in an increase in the number of moles of gas generally shows an increase in entropy.
   b. A reaction that results in a decrease in the number of moles of gas generally shows a decrease in entropy.
4. ΔS° and temperature
   ΔS° is nearly independent of temperature. The individual standard molar entropies (S°) of each reactant and product increase with temperature (because movement increases), but the *difference* in entropy ΔS° does not change appreciably with temperature.
5. ΔS° versus pressure
   The entropy change is not independent of pressure when gases are involved. Hence, in all calculations in this chapter it will be assumed that pressure is kept at 1 atm.
6. ΔS° versus concentration
   The entropy change is not independent of concentration when ions are involved. Again, for all calculations in this chapter it will be assumed that any ions present have a concentration of 1 M.
7. *Exercises*
   a. Calculate ΔH° and ΔS° for the reaction between copper, hydrogen ions, and nitrate ions to produce copper(II) ions, nitrogen dioxide gas, and water. *Solution:* The equation for the reaction is

$$Cu\ (s) + 2\,NO_3^-\ (aq) + 4\,H^+\ (aq) \rightarrow Cu^{2+}\ (aq) + 2\,NO_2\ (g) + 2\,H_2O$$

Note that you will not be able to do the calculations correctly unless you first balance the equation.

To get $\Delta H°$ we use the formula

$$\Delta H° = \sum \Delta H_f° \text{ products} - \sum \Delta H_f° \text{ reactants}$$
$$= [(64.8) + 2(33.2) + 2(-285.8)] - [2(-205.0)] = -30.4 \text{ kJ}$$

Remember that $\Delta H_f°$ is in kJ/mol, but $\Delta H°$ is in kJ.
$\Delta S°$ can be obtained similarly. To get $\Delta S°$ we use the formula

$$\Delta S° = \sum S° \text{ products} - \sum S° \text{ reactants}$$
$$= [(-99.6) + 2(240.0) + 2(69.9)] - [2(146.4) + (33.2)]$$
$$= 194.2 \frac{J}{K}$$

b. Calculate $\Delta H_f°$ and $S°$ for $CaSO_3 \cdot 2H_2O$ given the following information: **(E4)**

$$Ca(s) + SO_3(g) + 2H_2O(\ell) \rightarrow CaSO_3 \cdot 2H_2O(s)$$

$$\Delta H° = -795 \text{ kJ} \qquad \Delta S° = -253.5 \frac{J}{K}$$

D. Second law of thermodynamics

Statement of the law: _____

_____

or

$\Delta S_{univ} = $ _____ + _____ ____ 0

III. Free Energy Change (ΔG)
A. ΔG and spontaneity
1. If $\Delta G < 0$, the reaction is _____ .
2. If $\Delta G > 0$, the reaction is _____ .
3. If $\Delta G = 0$, the reaction is _____ .
4. Note that spontaneity can be related to the sign of ΔG only if the reaction is carried out at constant temperature and pressure.
B. Relation among ΔG, ΔH, and ΔS
1. Gibbs-Helmholtz equation:

$$\Delta G = \Delta H - T\Delta S$$

This equation is valid under all conditions of temperature, pressure, and concentration.
2. Two factors that tend to make ΔG negative and hence lead to a spontaneous reaction
   a. A negative value of ΔH
      (1) Exothermic reactions *tend* to be spontaneous. (This is not a universal statement.)
      (2) On the molecular level, there is a tendency to form "strong" bonds at the expense of "weak" ones.
   b. A positive value of ΔS
      There is a tendency for a reaction to be spontaneous if the products are less ordered than the reactants.

IV. Standard Free Energy Change (ΔG°)
A. Gibbs-Helmholtz equation
1. We have restricted ΔS to be ΔS°, which is the entropy change for reactions at 1 atm pressure and 1 M concentration for ions in solution. The Gibbs-Helmholtz equation takes the form

$$\Delta G^\circ = \Delta H^\circ - T\Delta S^\circ$$

where ΔG° is the standard free energy at 1 atm pressure and 1 M concentration for ions in solution.
2. *Exercises*
   a. Calculate ΔG° at 25°C for the reaction

$$Cu\,(s) + 4\,H^+\,(aq) + 2\,NO_3^-\,(aq) \rightarrow Cu^{2+}\,(aq) + 2\,NO\,(g) + 2\,H_2O\,(\ell)$$

Solution: We use the Gibbs-Helmholtz equation

$$\Delta G^\circ = \Delta H^\circ - T\Delta S^\circ$$

Since we have already calculated $\Delta H°$ to be -30.4 kJ and $\Delta S°$ to be 194.2 J/K, we can use these values here. Remember that we have to change $\Delta S°$ into kJ/K so our units will come out right. Also T has to be in Kelvin. Substituting then, we get

$$\Delta G° = -30.4 \text{ kJ} - \left(0.1942\frac{\text{kJ}}{\text{K}} \times 298\text{ K}\right) = -88.3 \text{ kJ}$$

The reaction is spontaneous because $\Delta G° < 0$.

   b. Show by calculation whether the dissolving of ammonium nitrate in water at 25°C
     (1) is exothermic or endothermic.
     (2) increases or decreases in entropy.
     (3) is spontaneous. **(E5)**

B. Standard free energy of formation

   1. Symbol: _____

   2. Definition: _____

        _____

   3. There are two ways of calculating the standard free energy of formation:
     a. $\Delta G°$ at T is equal to $\Sigma \Delta G_f°$ products at T minus $\Sigma \Delta G_f°$ reactants at T. This method is convenient if you have a table for the standard free energies of formation ($\Delta G_f°$) of compounds and ions at the temperature at which you want to calculate $\Delta G°$. It means that you will need a separate table for each temperature at which you might want to calculate $\Delta G°$.
     b. The Gibbs-Helmholtz equation.
        You can use $\Delta H_f°$ of the compounds and ions from the tables in your text. For $\Delta S_f°$, you can write out a reaction forming the compound from its elements. Use this reaction and Table 17.1 of your text to calculate $\Delta S_f°$. Plug both these values into the Gibbs-Helmholtz equation. Use the temperature at which you want to calculate $\Delta G_f°$ for T T.

   4. *Exercises*
     a. Calculate $\Delta G_f°$ for Fe_2O_3 at 35°C.
       Solution: We cannot use the table in the appendix of your text because that is a table for $\Delta G_f°$ at 25°C. Hence, we use the Gibbs-Helmholtz equation

$$\Delta G° = \Delta H° - T\Delta S°$$

where $\Delta H_f^\circ = -824.2$ kJ/mol. For ΔS_f°, we need to write out an equation for the formation of one mole of Fe_2O_3 from its elements:

$$4\,Fe\,(s) + 3\,O_2\,(g) \rightarrow 2\,Fe_2O_3\,(s)$$

We now calculate ΔS_f°.

$$\Delta S_f^\circ = [2(87.4)] - [4(27.3) + 3(205.0)] = -549.4 \text{ J/K}$$

The reaction shows that 2 moles of Fe_2O_3 are produced. Since ΔG_f° is in terms of kJ/mol, we need to find ΔS_f° for one mole of Fe_2O_3 instead of two. Dividing the value we get above by 2, we obtain ΔS_f° for one mole of Fe_2O_3: -274.7 J/mol·K. We now substitute both values into the Gibbs-Helmholtz equation

$$\Delta G_f^\circ = -824.2\,\frac{kJ}{mol} - 308\,K \times \left(-0.2747\,\frac{kJ}{mol \cdot K}\right)$$

$$= -739.6\,\frac{kJ}{mol}$$

b. Calculate the temperature at which ΔG_f° for HBr (g) is -60.0 kJ/mol. **(E6)**

V. Effect of Temperature, Pressure and Concentration on Spontaneity
A. Temperature
1. Study Table 17.2 of your text. You do not really need to memorize the table. You can construct it by analyzing the effect of the T change on the different signs in the Gibbs-Helmholtz equation.
2. You may be asked to calculate the temperature at which a reaction becomes spontaneous. Follow these steps:
   a. Calculate ΔH° and ΔS°.
   b. Check the signs of ΔH° and ΔS°.
   c. If $\Delta H^\circ < 0$ and $\Delta S^\circ > 0$, you do not have to go any further. The reaction is spontaneous at any temperature.
   d. If $\Delta H^\circ > 0$ and $\Delta S^\circ < 0$, stop. The reaction will never be spontaneous.
   e. If ΔH° and ΔS° have the same sign, calculate T when $\Delta G^\circ = 0$ (i.e., at equilibrium).
   If ΔH° and ΔS° are both positive, then the reaction is spontaneous at a temperature *above* the equilibrium temperature.
   If ΔH° and ΔS° are both negative, then the reaction is spontaneous at a temperature *below* the equilibrium temperature.

3. Exercises

    a. Show by calculation whether the reaction

$$2\,NO_2\,(g) \;\rightarrow\; 2\,NO\,(g) + O_2\,(g)$$

is spontaneous at 25°C. If it is not, then calculate the temperature at which spontaneity occurs.

    Solution: We first have to calculate $\Delta H°$ and $\Delta S°$.

$$\Delta H° \;=\; [2(90.2)] - [2(33.2)] \;=\; 114.0 \text{ kJ}$$

$$\Delta S° \;=\; [2(210.7) + (205.0)] - [2(240.0)] \;=\; 146.4 \,\frac{J}{K}$$

Now we calculate $\Delta G°$ at 25°C.

$$\Delta G° \;=\; 114.0 - (298)(0.1464) \;=\; 70.4 \text{ kJ}$$

Thus, the reaction is nonspontaneous at 25°C since $\Delta G° > 0$. To answer the next question, we follow the steps previously outlined:

Step (a) $\Delta H° = 114.0$ kJ and $\Delta S° = 146.4$ J/K.

Step (b) $\Delta H° > 0$ and $\Delta S° >$, so we go to Step (e).

Step (e) Calculate T at equilibrium, that is, when $\Delta G°$ is 0.

$$0 \;=\; 114.0 - T(0.1464)$$

$$T \;=\; 779\,K$$

Since $\Delta H°$ and $\Delta S°$ are both positive, the temperature at which the reaction becomes spontaneous is above 779 K.

    b. Calculate the temperature (if any) at which the reaction

$$4\,Fe_3O_4\,(s) + O_2\,(g) \;\rightarrow\; 6\,Fe_2O_3\,(s)$$

becomes spontaneous. **(E7)**

B. Pressure and concentration

   1. There is a general relation for the free energy change that is valid under any conditions. It is

$$\Delta G = \Delta G° + RT \ln Q$$

where Q is the "reaction quotient" referred to in Chapter 12 and has the same mathematical form as K.

   2. Units in calculating ΔG:

     a. T is in kelvin.

     b. R is in kJ/K   (8.31×10^{-3} kJ/K).

     c. For the Q expression:

       (1) Gases enter as their partial pressures in atmospheres.

       (2) Species in aqueous solution enter in M.

       (3) Pure solids or liquids do not appear.

       (4) The solvent in a dilute solution does not appear.

  3. Exercises

    a. Consider the reaction between sodium and water.

$$2\,Na\,(s) + 2\,H_2O\,(\ell) \;\rightarrow\; 2\,Na^+\,(aq) + H_2\,(g) + 2\,OH^-\,(aq)$$

      (1) Write the expression for Q.

        Solution:

$$Q = [Na^+]^2 (P_{H_2}) [OH^-]^2$$

Note that the ions are enclosed in brackets to indicate that concentration in M for these species has to be entered into the expression. Note further that Na(s) and H_2O are not included because sodium is a solid and water is a pure liquid.

      (2) Calculate $\Delta G°$ at 25°C.

        Solution: First we find $\Delta H°$.

$$\begin{aligned}
\Delta H° &= [2(\Delta H_f° \; Na^+) + 2(\Delta H_f° \; OH^-)] - [2(\Delta H_f° \; H_2O\,(\ell))] \\
&= [2(-240.1) + 2(-230.0)] - [2(-285.8)] \\
&= -368.6 \text{ kJ}
\end{aligned}$$

Next we find $\Delta S°$

$$\Delta S° = [2(S° \; Na^+) + (S° \; H_2) + 2(S° \; OH^-)]$$

$$- [2(S° \; Na) + 2(S° \; H_2O\,(\ell))]$$

Note that we have to look up $S°$ for both Na and H_2. Its value for these two elements is not zero.

$$= [2(59.0) + (130.6) + 2(-10.8)] - [2(51.2) + 2(69.9)]$$

$$= -15.2 \; \frac{J}{K}$$

We use the Gibbs-Helmholtz equation to find $\Delta G°$.

$$\Delta G° = \Delta H° - T\Delta S°$$
$$= -368.6 - 298(-0.0152) = -364.1 \text{ kJ}$$

(3) Find ΔG when $P_{H_2} = 750$ mm Hg, $[Na^+] = 0.10$ M, and the pH is 9.56.
Solution: We first find $[OH^-]$ from pH.

$$pOH = 14.00 - 9.56 = 4.44$$
$$[OH^-] = 3.6 \times 10^{-5} \text{ M}$$

Now we substitute the values given and obtained into the equation

$$\Delta G = \Delta G° + RT \ln Q$$

That gives

$$\Delta G = -364.1 + (8.31 \times 10^{-3})(298)\ln\left[(0.10)^2 \left(\frac{750}{760}\right)(3.6 \times 10^{-5})^2\right]$$

$$= -364.1 + (8.31 \times 10^{-3})(298)(-25.1)$$

$$= -426.2 \text{ kJ}$$

b. Consider the unbalanced equation

$$H^+ (aq) + MnO_2 (s) + Cl^- (aq) \rightarrow Mn^{2+} (aq) + Cl_2 (g)$$

(1) Write a balanced equation for the reaction.
(2) Calculate $\Delta G°$ at 40°C.
(3) Find ΔG at 40°C when the pH is 3.50, Cl_2 has a partial pressure of 0.950 atm, and the ionic species are at 0.100 M. **(E8)**

VI. The Free Energy Change and the Equilibrium Constant

A. Mathematical relationship

$$\Delta G° = -RT \ln K$$

Remember: K stands for any equilibrium constant used in the different types of equilibria. R is the gas constant 8.31 J/K and $\Delta G°$ has to be expressed in joules.

B. K versus the sign of G.

1. If $\Delta G° < 0$, then $\ln K > 0$; $K > 1$.
This means that when all species are at unit concentrations the reaction will proceed spontaneously in the left to right direction.

2. If $\Delta G° > 0$, then $\ln K < 0$; $K < 1$.
This means that when all species are at unit concentrations the reaction will proceed spontaneously in the right to left direction.

3. If $\Delta G° = 0$, then $\ln K = 0$; $K = 1$.
This means that the reaction is at equilibrium.

C. *Exercises*

1. Calculate K_{sp} for $PbCl_2$ at 75°C using the thermodynamic data available in the appendix of your text. Compare with its K_{sp} at 25°C.

Solution: In order to calculate for K_{sp} using thermodynamic data, we first need to know $\Delta G°$ at 75°C. To find $\Delta G°$, we in turn need $\Delta H°$ and $\Delta S°$ for the reaction

$$PbCl_2 (s) \rightleftharpoons Pb^{2+} (aq) + 2\,Cl^- (aq)$$

$$\Delta H° = -1.7 + 2(-167.2) - (-359.4) = 23.3 \text{ kJ}$$

$$\Delta S° = 10.5 + 2(56.5) - (136.0) = -12.5 \frac{J}{K}$$

$$\Delta G° = 23.3 - 348(-0.0125) = 27.6 \text{ kJ}$$

We now substitute $\Delta G°$ into the formula

$$\Delta G° = -RT \ln K$$

$$\ln K = \frac{-27.6 \times 10^3 \text{ J}}{8.31 \frac{J}{K} \times 348 \text{ K}} = -9.56$$

$$K_{sp} = 7.0 \times 10^{-5}$$

At 25°C, $K_{sp} = 1.7 \times 10^{-5}$, which means that $PbCl_2$ is more soluble in water at 75°C than at 25°C.

2. Calculate $\Delta G°$ for the dissolving of silver acetate ($AgC_2H_3O_2$) to make a 1 M solution at 25°C. Is the reaction spontaneous at this temperature? **(E9)**

VII. Additivity of Free Energy Changes; Coupled Reactions

A. Free energy changes are additive. If

$$\text{Reaction } 3 = \text{Reaction } 1 + \text{Reaction } 2$$

then

$$\Delta G_3 = \Delta G_1 + \Delta G_2$$

B. Since free energy changes are additive, it is possible to bring about a nonspontaneous reaction by coupling it with a reaction for which $\Delta G°$ is a large negative number.

SELF-TEST
A. Multiple Choice:

1. For a certain reaction, the standard free energy change is −70.0 kJ at 100 K and −40.0 kJ at 200 K. For this reaction
   a. $\Delta H > 0$; $\Delta S > 0$
   b. $\Delta H > 0$; $\Delta S < 0$
   c. $\Delta H < 0$; $\Delta S > 0$
   d. $\Delta H < 0$; $\Delta S < 0$
   e. need more information

2. How many of the following statements are true at 25°C and 1 atm?

   — $\Delta H_f°$ of H_2 (ℓ) = 0.0 kJ
   — $\Delta H_f°$ of H^+ (aq) = 0.0 kJ
   — $S°$ of H_2 (g) = 0.0 kJ
   — $S°$ of H^+ (aq) = 0.0 kJ

   a. 0   **b.** 1   **c.** 2   **d.** 3   **e.** 4

3. If the value of $\Delta G°$ does not change with temperature and pressure, this would mean that
   a. ΔH has a positive or negative value.   **b.** $\Delta H = 0$
   c. ΔS has a positive or negative value.   **d.** $\Delta S = 0$

4. Consider the following chemical processes.

   (1) $2 H_2$ (g) + O_2 (g) → $2 H_2O$ (g)
   (2) C (s) + H_2O (ℓ) → CO (g) + H_2 (g)
   (3) $3 Fe$ (s) + $4 H_2O$ (g) → $4 H_2$ (g) + Fe_3O_4 (s)
   (4) $2 Ag$ (s) + Cl_2 (g) → $2 AgCl$ (s)
   (5) $2 NO_2$ (g) → $2 O_2$ (g) + N_2 (g)

   The entropy change, $\Delta S°$ is expected to be positive in
   a. (1),(2)   **b.** (1),(3)   **c.** (2),(3)   **d.** (2),(4)   **e.** (2),(5)

5. For a dissociation reaction
    a. $\Delta S < 0$ **b.** $\Delta S = 0$ **c.** $\Delta S > 0$
    d. ΔS depends on the equilibrium constants.
    e. none of the above is true.

6. Which of the following statements is *always* true?
    a. An exothermic reaction is spontaneous.
    b. A reaction for which $\Delta S°$ is positive is spontaneous.
    c. If the number of moles of gas does not change in a chemical reaction, then
       $\Delta S° = 0$.
    d. If $\Delta H°$ and $\Delta S°$ are both positive, then $\Delta G°$ will decrease when the temperature
       increases.

7. For the sublimation of dry ice,

$$CO_2 \; (s) \;\; \rightarrow \;\; CO_2 \; (g)$$

the signs of $\Delta H°$ and $\Delta S°$ should be
    a. $\Delta H° > 0$; $\Delta S° > 0$ **b.** $\Delta H° < 0$; $\Delta S° < 0$
    c. $\Delta H° > 0$; $\Delta S° < 0$ **d.** $\Delta H° < 0$; $\Delta S° > 0$
    e. need more information

8. Increasing the temperature at which a reaction takes place will have the greatest
    effect on
    a. $\Delta G°$ **b.** $\Delta S°$ **c.** $\Delta H°$ **d.** $\Delta H°/\Delta S°$

9. As observed in the real world about us, each of the following processes is sponta-
    neous except
    a. dissolving salt in water.
    b. rusting of an iron nail.
    c. formation of the elements from boiling water.
    d. combustion of gasoline.

10. Pick the best statement.
    a. The free energy change is independent of temperature.
    b. The entropy of a substance is independent of temperature.
    c. The entropy change of a reaction is independent of temperature.
    d. The enthalpy of a substance is independent of temperature.

B. True or False:

_____ **1.** The second law of thermodynamics states that in a spontaneous process there is a net increase in entropy, taking into account the system and the surroundings.

_____ **2.** Energy is conserved in chemical reactions.

_____ **3.** The enthalpy change for spontaneous reactions is always negative.

_____ **4.** The free energy change for a reaction that increases in entropy is always positive.

_____ **5.** If ΔG is 0, then $\Delta S = \Delta H$.

_____ **6.** ΔS for a reaction between two gases to produce a liquid is negative.

_____ **7.** When a liquid freezes, its entropy increases.

_____ **8.** Spontaneous reactions always occur rapidly.

_____ **9.** As is true with enthalpies, absolute entropy values cannot be determined.

_____ **10.** At 0 K, a perfect crystal has $S = 0$.

C. Less than, Equal to, Greater than

Answer the questions below, using **LT** (for _is less than_), **GT** (for _is greater than_), **EQ** (for _is equal to_), or **MI** (for _more information required_) in the blanks provided.

The reaction given below takes place in a cylinder which feels warm to the touch after the reaction is complete.

$$A_2 (g) + B_2 (g) \rightarrow 2\,AB\,(s)$$

_____ **1.** At all temperatures, $\Delta S°$ ___ 0.

_____ **2.** At all temperatures, $\Delta H°$ ___ 0.

_____ **3.** At all temperatures, $\Delta G°$ ___ 0.

_____ **4.** $S°$ for AB (ℓ) ___ $S°$ for AB (s).

_____ **5.** $\Delta H_f°$ for A_2 (g) at 25°C and 1 atm ___ 0.

D. Problems:

Consider methanol (CH_3OH).

1. Write the equation for the combustion of two moles of methanol with oxygen into carbon dioxide and liquid water.

2. Calculate $\Delta H°$ for the reaction.

3. Calculate $\Delta S°$ for the reaction.

4. Calculate $\Delta G°$ for the reaction at 25°C.

5. Calculate $\Delta G°$ for the reaction at 100°C.

6. At which temperature (if any) will the reaction be spontaneous?

7. Calculate the boiling point of methanol if $\Delta H^\circ_{vap} = 37.95$ kJ and $\Delta S^\circ_{vap} = 112.9$ J/K.

8. Calculate ΔG at 25°C when $P_{CO_2} = 730$ mm Hg and $P_{O_2} = 1.1$ atm.

ANSWERS
Exercises:
(E1) The house being built

(E2) Before the game

(E3) The team after the huddle

(E4) $\Delta H_f^\circ = -1762$ kJ/mol; $S^\circ = 184.4$ J/K

(E5) $\Delta H^\circ = 28.1$ kJ; reaction endothermic; $S^\circ = 108.7$ J/K; increase in entropy; $\Delta G^\circ = -4.3$ kJ; reaction spontaneous

(E6) 413 K

(E7) any temperature < 1772 K

(E8) (1) $4\,H^+$ (aq) $+ MnO_2$ (s) $+ 2\,Cl^-$ (aq) $\rightarrow$ Mn^{2+} (aq) $+ 2\,H_2O + Cl_2$ (g)
(2) 23.4 kJ   **(3)** 113 kJ

(E9) $\Delta G^\circ = 16$ kJ;   nonspontaneous

Self-Test
A. Multiple Choice:
1. d	**2.** c	**3.** d	**4.** e	**5.** c
6. d	**7.** a	**8.** a	**9.** c	**10.** c

B. True or False:
1. T	**2.** T	**3.** F	**4.** F	**5.** F
6. F	**7.** F	**8.** F	**9.** F	**10.** T

C. Less than, Equal to, Greater than:
1. LT   **2.** LT   **3.** MI   **4.** GT   **5.** MI

D. Problems:
1. $2\,CH_3OH$ (ℓ) $+ 3\,O_2$ (g) $\rightarrow$ $2\,CO_2$ (g) $+ 4\,H_2O$ (ℓ)

2. $\Delta H^\circ = -1452.8$ kJ

3. $\Delta S^\circ = -161.8$ J/K

4. $\Delta G^\circ = -1404.6$ kJ

5. $\Delta G^\circ = -1392.4$ kJ

6. any temperature below 8979 K

7. 336.1 K

8. -1405.5 kJ

18

Electrochemistry

I. The Voltaic Cell
   A. How does it work?
   It produces electrical energy by carrying out a spontaneous redox reaction.
   B. Components
   1. Cathode
      a. Where the ion or molecule undergoes _____

      b. The ions that move toward the cathode are called _____.
   2. Anode
      a. Where the ion or molecule undergoes _____

      b. The ions that move toward the anode are called _____.

      c. Electrons are formed at the _____ and move toward the

      _____.

   3. Salt bridge

      a. Function: _____

      _____

      b. Brief description: _____

      _____

   C. Notation
   1. A voltaic cell can be abbreviated in the following manner:

   anode reaction ‖ cathode reaction

   2. The salt bridge is indicated by the symbol ‖.
   3. The anode and cathode reactions are further abbreviated

   reactant | product

4. The voltaic cell

$$Sn\ (s) + Zn^{2+}\ (aq) \rightarrow Zn\ (s) + Sn^{2+}\ (aq)$$

is thus abbreviated

$$Sn\ |\ Sn^{2+}\ \|\ Zn^{2+}\ |\ Zn$$

5. If no metal is involved in either anode or cathode reaction, or another metal is used as an electrode, the electrode is separated from the reactant|product entry. If the electrode is used as the anode then the symbol for the metal is written before the reactant for the anode and separated by |. If the electrode is used as the cathode then the symbol for the metal is written after the product for the cathode and separated by |.

D. Drawing a diagram of the cell

Key points to remember:

1. There are always two separate half-cells connected by a wire and a salt bridge.

2. One of the compartments is the anode and the other is the cathode.
   All species (reactants and products of a half-reaction), except water, are shown in the appropriate compartment.

3. If a metal precipitates in a cell half-reaction (either as product or reactant), it is ordinarily chosen as the electrode for the compartment where that half-reaction occurs.
   If no metal is involved in the half-reaction, an electrically conducting non-reactive solid like platinum (Pt) or graphite (C) should be used.

4. Electron flow from anode to cathode is denoted by an arrow in the external circuit.

5. The flow of ions through the salt bridge is also indicated by an arrow. Cations move to the cathode, anions move to the anode.

E. *Exercises*

1. When potassium permanganate and sodium iodide are mixed, a spontaneous reaction occurs:

$$2\ MnO_4^-\ (aq) + 10\ I^-\ (aq) + 16\ H^+\ (aq) \rightarrow 5\ I_2\ (aq) + 2\ Mn^{2+}\ (aq) + 8\ H_2O$$

This reaction can serve as a source of energy in a voltaic cell.

a. Write the anode and cathode reactions.

b. Draw a diagram of the cell. Indicate which way the electrons move in the external circuit.

c. Write the cell notation.

Solution:

a. Since Mn goes from an oxidation number of 7 to 2, it is being reduced. I goes from an oxidation number of -1 to 0, so it is oxidized. Thus the half-cell reactions are:

anode: $2\ I^-\ (aq) \rightarrow I_2\ (aq) + 2\ e^-$

cathode: $MnO_4^-\ (aq) + 8\ H^+\ (aq) + 5\ e^- \rightarrow Mn^{2+}\ (aq) + 4\ H_2O$

b. The diagram of the cell is

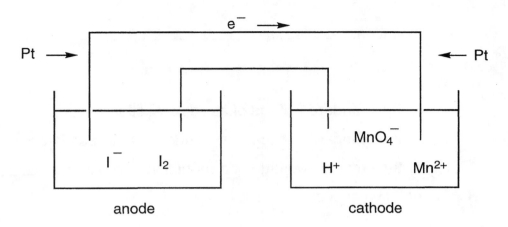

anode   cathode

c. Cell notation: $Pt \mid I^- \mid I_2 \parallel MnO_4^- \mid Mn^{2+} \mid Pt$

2. Given the cell notation

$$Fe \mid Fe^{2+} \parallel MnO_2 \mid Mn^{2+} \mid Pt$$

a. Write the reaction that takes place at the anode; at the cathode. Write the overall cell reaction. **(E1)**

b. Draw a diagram of the cell, indicating the anode, the cathode, all species in each half-cell and the direction of electron flow in the external circuit. **(E2)**

II. **Standard Voltages**
   A. Qualitative description
      1. Definition: _____

         _____

      2. Symbol for
         a. the standard voltage of a redox reaction: _____
         b. the standard voltage of an oxidation half-reaction: _____
         c. the standard voltage of a reduction half-reaction: _____

   B. Quantitative aspects
      1. $E° = E°_{ox} + E°_{red}$
      2. $E°_{red}$ for the following reaction

         $$2\,H^+\,(aq\,,\,1\,M) + 2\,e^- \;\rightarrow\; H_2\,(g\,,\,1\,atm)$$

      is zero.
   C. Standard potentials
      1. The standard potentials listed in Table 18.1 of the text give you the standard voltages for *reduction* half-reactions.
         Remember: These are half-reactions in which e^- are on the reactant side and there is a reduction in oxidation number.
      2. Standard voltages for oxidation half-reactions are obtained by changing the *sign* of the reduction half-reaction. You can do this only if the oxidation half-reaction is the reverse of the reduction half-reaction as written in Table 18.1.
   D. The strength of oxidizing and reducing agents
      1. The larger $E°_{red}$ is, the stronger the oxidizing agent is. Remember the oxidizing agent is the *reactant* in a reduction half-reaction.
      2. The larger $E°_{ox}$ is, the stronger the reducing agent is.
      3. In order to compare the strength of reducing agents, you must:
         a. Find the species on the right side of the standard potential table. If the ion or molecule is on the left side, it is not acting as a reducing agent.
         b. Reverse the reactions, making sure to change $E°_{red}$ to $E°_{ox}$ and reverse the sign.
         c. Compare the magnitude of $E°_{ox}$ values.
      4. In order to compare the strength of oxidizing agents, you must:
         a. Find the species on the left-hand side of the standard potential table. If the ion or molecule is on the right-hand side, it is not acting as an oxidizing agent.
         b. Compare the magnitude of $E°_{ox}$ values.
      5. The product formed by a strong oxidizing agent is itself a weak reducing agent.
      6. The product formed by a strong reducing agent is itself a weak oxidizing agent.

7. *Exercises*

    a. Arrange, in decreasing order of strength, the following oxidizing agents in acidic solution.

$$I_2 \text{ (s)}; \quad ClO_3^- \text{ (aq)}; \quad F_2 \text{ (g)}; \quad Na^+ \text{ (aq)}; \quad Zn^{2+} \text{ (aq)}; \quad PbO_2 \text{ (s)}$$

Solution: Since we are ranking oxidizing agents, these are the species that are reduced. Hence, we should find them on the left side of the equations listed in Table 18.1. The equations are

$I_2 \text{ (s)} + 2\,e^- \rightarrow 2\,I^- \text{ (aq)}$	$E_{red}^\circ = 0.534 \text{ V}$
$ClO_3^- \text{ (aq)} + 6\,H^+ \text{ (aq)} + 5\,e^- \rightarrow \frac{1}{2}\,Cl_2 \text{ (g)} + 3\,H_2O$	$E_{red}^\circ = 1.458 \text{ V}$
$F_2 \text{ (g)} + 2\,e^- \rightarrow 2\,F^- \text{ (aq)}$	$E_{red}^\circ = 2.889 \text{ V}$
$Na^+ \text{ (aq)} + e^- \rightarrow Na \text{ (s)}$	$E_{red}^\circ = -2.714 \text{ V}$
$Zn^{2+} \text{ (aq)} + 2\,e^- \rightarrow Zn \text{ (s)}$	$E_{red}^\circ = -0.762 \text{ V}$
$PbO_2 \text{ (s)} + SO_4^{2-} \text{ (aq)} + 4\,H^+ \text{ (aq)} + 2\,e^- \rightarrow$ $\qquad PbSO_4 \text{ (s)} + 2\,H_2O$	$E_{red}^\circ = 1.687 \text{ V}$

From this table we see that the largest E_{red}° is 2.889, which belongs to the reduction half-reaction of F_2. Thus, F_2 is the strongest oxidizing agent. We go through the same reasoning for all other oxidizing agents and get the order

$$F_2 > PbO_2 > ClO_3^- > I_2 > Zn^{2+} > Na^+$$

    b. Arrange in increasing order of strength the following reducing agents in acidic solution. **(E3)**

$$Cu^+ \text{ (aq)} \quad Co \text{ (s)} \quad NO \text{ (g)} \quad Co^{2+} \text{ (aq)} \quad K \text{ (s)}$$

E. Calculation of E° from E_{red}° and E_{ox}°
    1. $E^\circ = E_{red}^\circ + E_{ox}^\circ$
    2. When given a redox equation, you have to break it up into two half-equations.
        a. The reduction half-reaction has voltage E_{red}°. You get this half-reaction "as is" from Table 18.1.
        b. The oxidation half-reaction has voltage E_{ox}°. You have to look for the reverse of that reaction in Table 18.1. When you have found it, write your oxidation

half-reaction, and take the voltage of the reverse reaction. Do not forget to reverse the sign.

3. *Exercises*

  a. Calculate $E°$ for the reaction

$$Cu^{2+} (aq) + H_2S (g) \rightarrow S (s) + 2 H^+ (aq) + Cu (s)$$

Solution: In order to decide how to split the reaction, we have to determine oxidation numbers.

$$Cu : +2 \text{ to } 0; \quad H : +1 \text{ to } +1; \quad S : -2 \text{ to } 0$$

Thus, Cu is reduced and H_2S is oxidized.
Reduction half-reaction:

$$Cu^{2+} (aq) + 2 e^- \rightarrow Cu (s) \qquad\qquad E°_{red} = 0.339 \text{ V}$$

Oxidation half-reaction:

$$H_2S (g) \rightarrow 2 H^+ (aq) + S (s) + 2 e^-$$

Looking this up in the table, we find the following equation

$$2 H^+ (aq) + S (s) + 2 e^- \rightarrow H_2S (g) \qquad\qquad E°_{red} = 0.144 \text{ V}$$

Since we want $E°_{ox}$ for the reverse reaction, we write

$$H_2S (g) \rightarrow 2 H^+ (aq) + S (s) + 2 e^- \qquad\qquad E°_{ox} = -0.144 \text{ V}$$

Thus,

$$E° = 0.339 + (-0.144) = 0.195 \text{ V}$$

  b. Calculate $E°$ for the cell

$$Pt \,|\, Cl^- (aq), \; 1 \text{ M}) \,|\, Cl_2 (g, \; 1 \text{ atm}) \,\|\, Cu^{2+} (aq, \; 1 \text{ atm}) \,|\, Cu (s) \qquad\qquad \textbf{(E4)}$$

F. Spontaneity of redox reactions
   1. If $E° > 0$, the reaction is spontaneous.
   2. If $E° < 0$, the reaction is nonspontaneous; the reverse reaction will tend to occur.
   3. If $E° = 0$, the reaction is at equilibrium.

III. Relations between E°, ΔG°, and K
   A. Indicators of spontaneity
      1. $E°$
         a. If $E° > 0$, the reaction is spontaneous.
         b. If $E° < 0$, the reaction is nonspontaneous.
         c. If $E° = 0$, the reaction is at equilibrium.
      2. $\Delta G°$
         a. If $\Delta G° > 0$, the reaction is nonspontaneous.
         b. If $\Delta G° < 0$, the reaction is spontaneous.
         c. If $\Delta G° = 0$, the reaction is at equilibrium.
      3. K
         a. If $K > 1$, the reaction occurs spontaneously from left to right at standard conditions.
         b. If $K < 1$, the reverse reaction occurs spontaneously.
   B. Relation between E° and ΔG°
      1. Mathematical relationship

$$\Delta G° = -nFE°$$

where n is the number of electrons transferred. F is the Faraday constant, 96,485. Thus, in kJ

$$\Delta G° = -96.5\,n\,E°$$

   2. *Exercises*
      a. Determine $\Delta G°$ for the equation

$$4\,H^+\,(aq) + 2\,Cl^-\,(aq) + MnO_2\,(s) \;\rightarrow\; Mn^{2+}\,(aq) + 2\,H_2O + Cl_2\,(g)$$

Solution:
(1) We first determine E° by writing the two half-reactions and their standard potentials. The oxidation half-reaction is

$$2\,Cl^-\,(aq) \;\rightarrow\; Cl_2\,(g) + 2\,e^- \qquad\qquad E°_{ox} = -1.360\ V$$

The reduction half-reaction is

$$2\,e^- + 4\,H^+\,(aq) + MnO_2\,(s) \;\rightarrow\; Mn^{2+}\,(aq) + 2\,H_2O$$
$$E°_{red} = 1.229\ V$$

Calculating E°, we get

$$E° = -1.360 + 1.229 = -0.131\ V$$

Since E° is negative, we know that the reaction is nonspontaneous and expect $\Delta G°$ to be positive.

(2) We note that two electrons are transferred, thus n = 2.

(3) To calculate $\Delta G°$, we substitute into the equation

$$\Delta G° = -96.5\, n\, E°$$

and get

$$\Delta G° = -96.5\,(2)\,(-0.131) = 25.3\,kJ$$

b. If $E°_{red}$ for the reduction of PbO_2 to Pb^{2+} is 1.455 V, calculate $\Delta G°$ for the unbalanced redox reaction   **(E5)**

$$Pb^{2+}\,(aq) + MnO_4^-\,(aq) + H_2O \;\rightarrow\; PbO_2\,(s) + Mn^{2+}\,(aq) + H^+\,(aq)$$

C. Relation between E° and K

1. Mathematical relationship at 25°C

$$\ln K = \frac{n\,E°}{0.0257}$$

2. *Exercises*

a. Write a balanced redox equation for the reaction between sulfide ion and oxygen gas to produce solid sulfur and hydroxide ion. Calculate K for the reaction at 25°C.

Solution: The oxidation half-reaction is

$$S^{2-}\,(aq) \;\rightarrow\; S\,(s) + 2\,e^- \qquad\qquad E°_{ox} = 0.445\,V$$

The reduction half-reaction is

$$O_2\,(g) + 2\,H_2O + 4\,e^- \;\rightarrow\; 4\,OH^-\,(aq) \qquad\qquad E°_{red} = 0.401\,V$$

E° therefore is 0.846 V. We multiply the oxidation half-reaction by 2 to get the same number of electrons lost and gained. Thus, n = 4. We next calculate K.

$$\ln K = \frac{4 \times 0.846}{0.0257} = 132$$

Taking the inverse ln of both sides, we obtain

$$K = inv\, \ln 132 = 1{:}53 \times 10^{57}$$

    b. Write a balanced redox equation for the reaction between chlorine gas and nitrogen oxide gas in acid solution producing chloride ion and nitrate ion. Calculate K for the reaction. **(E6)**

IV. Effect of Concentration Upon Voltage

  A. Qualitative aspects of concentration change

    1. A reaction becomes more spontaneous (voltage increases) if

      a. the concentration of the reactant _____.

      b. the concentration of the product _____.

    2. A reaction becomes less spontaneous (voltage decreases) if

      a. the concentration of the reactant _____.

      b. the concentration of the product _____.

  B. Nernst Equation

    1. Note that E°_{red}, E°_{ox}, and E° apply only at standard conditions, that is, 1 M for species in solution, and 1 atm for gases.

    2. Mathematical relationship between concentration, E°, and E, the voltage at a given concentration
For the reaction

$$a\,A\,(\ell)\,+\,b\,B\,(aq)\;\rightarrow\;c\,C\,(aq)\,+\,d\,D\,(g)$$

the voltage E, at given conditions, is calculated using the equation

$$E\,=\,E^\circ\,-\,\frac{0.0257}{n}\,\ln\frac{[C]^c\,(P_D)^d}{[B]^b}$$

where

$E\,=$ voltage at given concentration

$E^\circ\,=$ standard voltage

$n\,=$ number of moles of e^- transferred

3. The condition for species in solution is expressed in molarity, while the condition for gases is expressed in partial pressure (in atm).

4. Pure liquids and solids do not appear.

5. *Exercises calculating E*

   a. Calculate the voltage of the cell at 25°C for the reaction

   $$10\,Cl^- (aq) + 2\,MnO_4^- (aq) + 16\,H^+ (aq) \rightarrow$$
   $$5\,Cl_2 (g) + 2\,Mn^{2+} (aq) + 8\,H_2O$$

   The partial pressure of chlorine gas is 1.00 atm, the pH is 0.50, and the concentrations of all the ionic species except H^+ are 0.10 M.

   Solution: We need to use the Nernst equation to calculate E, since the conditions are no longer the standard conditions of 1 M for ionic species and 1 atm for gases.

   (1) First, we must calculate $E°$. We do that by writing the two half-reactions.

   $$2\,Cl^- (aq) \rightarrow Cl_2 (g) + 2\,e^- \qquad\qquad E_{ox}° = -1.360 \text{ V}$$
   $$5\,e^- + 8\,H^+ (aq) + MnO_4^- (aq) \rightarrow Mn^{2+} (aq) + 4\,H_2O$$
   $$E_{red}° = 1.512 \text{ V}$$

   Note: To calculate $E_{ox}°$ and $E_{red}°$, we do not pay any attention to the coefficients used to balance the species in the final reaction. Each half-reaction is balanced independently and looked up.

   $$E° = E_{ox}° + E_{red}°$$
   $$= -1.360 + 1.512 = 0.152 \text{ V}$$

   (2) Next, we determine n. To do so, we multiply both equations by integers to make the number of electrons gained and lost the same. In this case, we multiply the oxidation half-reaction by 5.

   $$10\,Cl^- (aq) \rightarrow 5\,Cl_2 (g) + 10\,e^-$$

   We multiply the reduction half-reaction by 2.

   $$10\,e^- + 16\,H^+ (aq) + 2\,MnO_4^- (aq) \rightarrow 2\,Mn^{2+} (aq) + 8\,H_2O$$

   Notice that the coefficients of the species are now just like those given for the cell equation. Since there are 10 electrons gained as well as lost, n = 10.

   It is important that you first find $E°$ before you determine n, so that you will not be tempted to multiply $E_{ox}°$ and $E_{red}°$ by the integers you use to make e^- lost equal to e^- gained.

   (3) Lastly, we determine the conditions for the ionic species in molar concentrations, and those for gases in atm pressure. All the ionic species except H^+ have a concentration of 0.10 M, that is, $[Cl^-] = [MnO_4^-] = [Mn^{2+}] = 0.10$ M. $[H^+]$ is expressed in terms of pH. Since the pH is 0.50, $[H^+] = 0.32$ M. Chlorine gas has pressure of 1.00 atm.

(4) Substituting, we get

$$E = 0.152 - \frac{0.0257}{10} \ln \frac{(P_{Cl_2})^5 [Mn^{2+}]^2}{[Cl^-]^{10} [MnO_4^-]^2 [H^+]^{16}}$$

$$= 0.152 - 0.00257 \ln \frac{(1.0)^5 (0.10)^2}{(0.10)^{10} (0.10)^2 (0.32)^{16}}$$

$$= 0.152 - 0.00257 \ln (8.3 \times 10^{17})$$

$$= 0.046 \, V$$

b. Calculate the voltage of the cell

$$Cu \,|\, Cu^{2+} \,(0.10\,M) \,\|\, NO_3^- \,(0.20\,M) \,|\, NO \,(1\,atm) \,|\, Pt$$

if the pH of the cell is 2.00. **(E7)**

6. *Exercises calculating concentration or partial pressure*
   a. Calculate the pH of the cell below with a voltage of 3.14 V and carried out in basic solution.

$$Ca \,|\, Ca^{2+} \,(0.10\,M) \,\|\, CrO_4^{2-} \,(0.10\,M) \,|\, Cr(OH)_3 \,|\, Pt$$

Solution:
(1) First, we have to write the oxidation half-reaction and reduction half-reaction to calculate E°. The oxidation half-reaction is

$$Ca \,(s) \;\rightarrow\; Ca^{2+} \,(aq) + 2\,e^- \qquad\qquad E^\circ_{ox} = \; 2.869 \, V$$

The reduction half-reaction is

$$3\,e^- + 4\,H_2O + CrO_4^{2-} \,(aq) \;\rightarrow\; Cr(OH)_3 \,(s) + 5\,OH^- \,(aq)$$
$$E^\circ_{red} = -0.12 \, V$$

E° is therefore

$$E^\circ = 2.869 + (-0.12) = 2.75 \, V$$

(2) We multiply the oxidation half-reaction by 3, the reduction half-reaction by 2 and see that there are 6 electrons gained and lost. The overall equation is

$$2\,CrO_4^{2-}\,(aq) + 3\,Ca\,(s) + 8\,H_2O \longrightarrow$$
$$3\,Ca^{2+}\,(aq) + 10\,OH^-\,(aq) + 2\,Cr(OH)_3\,(s)$$

(3) We write the Nernst equation for this reaction, substitute and get

$$E = E° - \frac{0.0257}{6}\,\ln\frac{[Ca^{2+}]^3\,[OH^-]^{10}}{[CrO_4^{2-}]^2}$$

$$3.14 = 2.75 - 0.00428\,\ln\frac{(0.10)^3\,[OH^-]^{10}}{(0.10)^2}$$

$$= 2.75 - 0.00428\,\ln[(0.10)\,[OH^-]^{10}]$$

At this point, resist the temptation to "simplify" to get

$$3.14 = 2.74\,\ln[(0.10)\,[OH^-]^{10}]$$

This is wrong! Instead, you have to do the following:

$$0.00428\,\ln[(0.10)\,[OH^-]^{10}] = 2.75 - 3.14$$

$$\ln[(0.10)\,[OH^-]^{10}] = \frac{-0.39}{0.00428} = -91$$

Taking the inverse ln of both sides,

$$[OH^-]^{10}\,(0.10) = 3.0 \times 10^{-40}$$
$$[OH^-]^{10} = 3.0 \times 10^{-39}$$
$$[OH^-] = 1.4 \times 10^{-4}$$

Since $[H^+][OH^-] = 1 \times 10^{-14}$, $[H^+] = 7 \times 10^{-11}$ and the pH of the cell is 10.1.

b. Consider the cell with reaction

$$Fe\,(s) + 2\,H^+\,(aq) \longrightarrow Fe^{2+}\,(aq) + H_2\,(g)$$

When the voltage of the cell is 0.420 V the pH is determined to be 1.30

and the partial pressure of H_2 is 1.0 atm. What is the concentration of Fe^{2+} in the solution? **(E8)**

V. Electrolytic Cells
A. Qualitative aspects
1. The reaction in this cell is nonspontaneous.
2. A battery (source of electrical energy) acts as an electron pump, pushing electrons into the cathode and removing them from the anode.
3. An oxidation-reduction carried out in an electrolytic cell is called electrolysis.

B. Cell Reactions in Aqueous Solutions
1. Possible reactions at the cathode
   a. The cation gets reduced to _____ .

   b. Water gets reduced to _____ .

   Half-reaction: _____
2. Possible reactions at the anode
   a. The anion gets oxidized to _____ .

   b. Water gets oxidized to _____ .

   Half-reaction: _____

C. Quantitative aspects of electrolysis
1. When considering a half-reaction as in
$$Ni^{2+} (aq) + 2\,e^- \;\rightarrow\; Ni\,(s)$$
you can write the following equivalences:
$$1 \text{ mol } Ni^{2+} = 2 \text{ mol } e^- = 1 \text{ mol } Ni$$
2. A mole of electrons is equal to 96,485 coulombs. This is known as the Faraday constant.
3. One coulomb is equal to 1 amp-s.
4. *Exercises*
   a. How many grams of copper metal can be obtained by passing 1.20 amp of electric current for half an hour through an aqueous solution of copper(II) ions?

Solution: First, we write the half-reaction

$$Cu^{2+} (aq) + 2e^- \rightarrow Cu (s)$$

Next we write the equivalences

$$1 \text{ mol } Cu^{2+} = 2 \text{ mol } e^- = 1 \text{ mol } Cu$$

Now, since we are given the number of amperes and how long the current goes through, we can calculate how many coulombs are involved. Remember that coulombs are ampere-seconds, hence time has to be expressed in seconds.

$$0.500 \text{ hr} \times \frac{60 \text{ min}}{1 \text{ hr}} \times \frac{60 \text{ s}}{1 \text{ min}} \times 1.20 \text{ amp} = 2.16 \times 10^3 \text{ amp} \cdot \text{s}$$

Since 2.16×10^3 amp·s is equal to 2160 coulombs, we can now convert coulombs (C) to moles of copper and then to grams of copper.

$$2.16 \times 10^3 \text{ C} \times \frac{1 \text{ mol } e^-}{96485 \text{ C}} \times \frac{1 \text{ mol } Cu}{2 \text{ mol } e^-} \times \frac{63.55 \text{ g}}{1 \text{ mol } Cu} = 0.711 \text{ g Cu}$$

b. How long will it take to deposit all the Ni^{2+} as Ni in 300.0 cm^3 of 0.2500 M of $NiSO_4$ with a current of 4.60 amp? **(E9)**

VI. Commercial Cells
A. Electrolysis of aqueous NaCl
   1. Reaction at the anode:

   2. Reaction at the cathode:

   3. Overall cell reaction:

B. Primary cells (non-rechargeable)
   1. Dry cell (Leclanche cell)

   a. The anode is _____ .

   b. The cathode is _____ .

   c. The space between electrodes is filled with a paste that contains

   _____ , _____ , and _____ .

    d. The half-reaction at the anode is

_____ .

    e. The half-reaction at the cathode is

_____ .

    f. The overall reaction is

_____ .

    g. In an alkaline cell _____ rather than _____ is used.
The overall reaction for an alkaline cell is

_____ .

  2. Mercury cell
    a. It is used in hearing aids, watches, cameras, etc.

    b. The anode is _____ .

       The reacting species is _____ .

    c. The cathode is _____ .

    d. The electrolyte is _____ .

    e. The half-reaction at the anode is _____ .

    f. The half-reaction at the cathode is _____ .
    g. The overall reaction is

_____ .

C. Storage (rechargeable) cells
  1. Lead storage battery
    a. The anode is made up of _____ .

    b. The cathode is made up of _____ .

    c. The anode and cathode are immersed in _____ .

    d. When a lead storage battery supplies current, it acts as a _____ .
The half-reaction at the anode is

_____ .

The half-reaction at the cathode is

_____ .

The overall reaction is is

_____ .

    e. When a lead storage battery is being recharged, it acts as a _____ .
The half-reaction at the anode is

_____ .

The half-reaction at the cathode is

_____ .

The overall reaction is

_____ .

2. Nicad battery

   a. The anode is _____ .

   b. The cathode contains _____ .

   c. The electrolyte is _____ .

   d. When the battery supplies energy,

     the half-reaction at the anode is

     _____ ;

     the half-reaction at the cathode is

     _____ ;

     the overall reaction is

     _____ .

SELF-TEST

A. Multiple Choice:

1. Using the table of standard reduction potentials, which of the following is the strongest oxidizing agent?

   a. Fe^{2+} **b.** Fe^{3+} **c.** Al^{3+} **d.** Cu^{2+}

2. The spontaneous reaction

$$Cd\,(s) + 2\,H^+\,(aq) \rightarrow Cd^{2+}\,(aq) + H_2\,(g)$$

is used as a source of energy in a voltaic cell. The electrode at the cathode would most likely be made of

   a. $Cd\,(s)$ **b.** $CdCl_2\,(s)$ **c.** $HCl\,(aq)$ **d.** $H_2\,(g)$ **e.** Pt metal

3. When a lead storage battery is used as a source of electrical energy

   a. $PbSO_4$ forms at the cathode.

   b. $PbSO_4$ forms at the anode.

   c. H^+ ions are consumed.

   d. all of the above are true.

   e. none of the above is true.

4. Which one of the following is capable of oxidizing Br^- ions to Br_2?
    a. Cl^- **b.** Zn   **c.** MnO_4^- **d.** H^+

5. For the reaction

$$3\,Cu\,(s) + 2\,NO_3^-\,(aq) + 8\,H^+\,(aq) \rightarrow 3\,Cu^{2+}\,(aq) + 2\,NO\,(g) + 4\,H_2O$$

E° is 0.44 V. How many of the following changes on the cell will affect the voltage?

    — increasing $[NO_3^-]$
    — increasing the pH
    — increasing $[Cu^{2+}]$
    — reducing P_{NO}

    a. 0   **b.** 1   **c.** 2   **d.** 3   **e.** 4

6. At standard conditions, a spontaneous reaction
    a. has $\Delta G^\circ < 0$.
    b. has $K > 1$.
    c. has $E^\circ > 0$.
    d. is characterized only by (a) and (c) above.
    e. involves (a), (b), and (c) above.

7. The anode in any electrochemical cell is

    (1) always positive.
    (2) on the side where oxidation occurs.
    (3) on the side where reduction occurs.
    (4) the electrode towards which the cations move.
    (5) the electrode towards which the anions move.

    a. (2),(5)   **b.** (3),(4)   **c.** (1),(2),(5)   **d.** (1),(3),(4)   **e.** (1),(2),(4)

8. If a chemical reaction is used in a voltaic cell, it must

    — be a redox type of reaction.
    — be spontaneous.
    — have $E^\circ > 0$.
    — have $K < 1$.
    — use the reactants as electrodes.

The number of true statements is
    a. 1   **b.** 2   **c.** 3   **d.** 4   **e.** 5

9. How many moles of electrons are transferred in the reaction as written?

$$3\,Ag\,(s) + NO_3^-\,(aq) + 4\,H^+\,(aq) \rightarrow 3\,Ag^+\,(aq) + NO\,(g) + 2\,H_2O$$

    a. 0   **b.** 1   **c.** 2   **d.** 3   **e.** 4

10. When a solution containing Cu^{2+} and Pb^{2+} is electrolyzed, Cu and PbO_2 are deposited on the electrodes. Which of the following statements is true?
   a. The Cu (s) and PbO_2 (s) are both deposited at the anode.
   b. The Cu (s) and PbO_2 (s) are both deposited at the cathode.
   c. Cu (s) forms at the anode and PbO_2 (s) forms at the cathode.
   d. Cu (s) forms at the cathode and PbO_2 (s) forms at the anode.

B. True or False:
 A. The standard hydrogen electrode potential is 0.00 V at 25°C because

_____ **1.** the voltmeter records 0.00 V for the half-cell.

_____ **2.** it has been measured precisely as 0.00 V with respect to many electrodes.

_____ **3.** it has been defined that way.

 B. A voltaic cell where $E° > 0$

_____ **4.** has $\Delta G° < 0$

_____ **5.** has K > 1.

_____ **6.** is always exothermic.

_____ **7.** may have E < 0.

C. Consider the standard reduction potential ($E°_{red}$) for chlorine gas
$$Cl_2 (g) + 2e^- \rightarrow 2Cl^- (aq) \qquad E°_{red} = 1.360\ V$$
   From this information, one can tell that

_____ **8.** chlorine gas is very difficult to oxidize.

_____ **9.** chlorine gas is an excellent reducing agent.

_____ **10.** $Cl_2 (g) + H_2 (g) \rightarrow 2H^+ (aq) + 2Cl^- (aq) \qquad \Delta G° < 0$

_____ **11.** $\frac{1}{2} Cl_2 (g) + e^- \rightarrow Cl^- (aq) \qquad E°_{red} = 0.680\ V$

_____ **12.** chloride ion is an excellent reducing agent.

C. Less than, Equal to, Greater than:

Answer the questions below, using **LT** (for *is less than*), **GT** (for *is greater than*), **EQ** (for *is equal to*), or **MI** (for *more information required*) in the blanks provided.

_____ 1. For the half reaction
$$2\,Al\,(s) \rightarrow 2\,Al^{3+}\,(aq) + 6\,e^-$$
E°_{ox} _____ 1.68 V.

_____ 2. For the reaction
$$2\,Al\,(s) + 3\,Ni^{2+}\,(aq) \rightarrow 2\,Al^{3+}\,(aq) + 3\,Ni\,(s)$$
E° for the cell is _____ 0 .

_____ 3. For the reaction given in (2), n is _____ 3.

_____ 4. For the reaction given in (2), the number of coulombs that passes through the cell is _____ $(6)(9.648 \times 10^4)$.

_____ 5. For the reaction given in (2), E is _____ E°.

D. Problems:

Consider the reaction between dichromate ions and bromide ions to give chromium(III) ions and liquid bromine.

1. Write the reduction half-reaction and find E°_{red}.

2. Write the oxidation half-reaction and find E°_{ox}.

3. Write the balanced redox reaction in acid solution and calculate E°.

4. Calculate E if all ionic species except H^+ have a concentration of 0.2 M and the solution has a pH of 1.5.

5. Calculate the pH if all the ionic species except H^+ have a concentration of 0.10 M and the voltage of the cell is 0.05 V.

6. Calculate $\Delta G°$ for the reaction at 25°C and standard conditions.

7. Calculate K for the reaction at 25°C and standard conditions.

Silver is electroplated from an aqueous solution of $AgClO_4$ in an electrolytic cell where a current of 5.00 amp is applied for 45.0 min. The following overall reaction takes place:

$$4\,Ag^+\,(aq)\,+\,2\,H_2O\,\rightarrow\,O_2\,(g)\,+\,4\,Ag\,(s)\,+\,4\,H^+\,(aq)$$

8. What mass of silver is produced by electrolysis?

9. What volume of oxygen at 1.0 atm pressure and 25°C is produced per gram of silver produced?

ANSWERS
Exercises:

(E1) cathode reaction: $MnO_2 (s) + 4 H^+ (aq) + 2 e^- \rightarrow Mn^{2+} (aq) + 2 H_2O$

anode reaction: $Fe (s) \rightarrow Fe^{2+} (aq) + 2 e^-$

overall reaction: $Fe (s) + MnO_2 (s) + 4 H^+ (aq) \rightarrow Fe^{2+} (aq) + Mn^{2+} (aq) + 2 H_2O$

(E2)

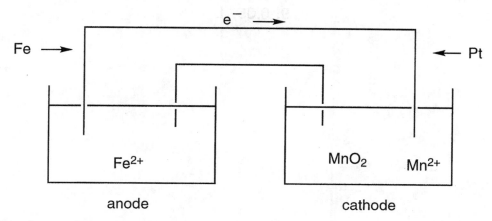

(E3) $Co^{2+} < NO < Cu^+ < Co < K$

(E4) $E° = -1.021$ V

(E5) -55 kJ

(E6) $3 Cl_2 (g) + 2 NO (g) + 4 H_2O \rightarrow 6 Cl^- (aq) + 2 NO_3^- (aq) + 8 H^+ (aq)$; 1.42×10^{40}

(E7) 0.5 V

(E8) 1.1×10^{-3}

(E9) 52.4 min

Self-Test
A. Multiple Choice:

1. b	**2.** e	**3.** d	**4.** c	**5.** e
6. e	**7.** a	**8.** c	**9.** d	**10.** d

B. True or False:

A. 1. F	**2.** F	**3.** T		
B. 4. T	**5.** T	**6.** F	**7.** T	
C. 8. T	**9.** F	**10.** T	**11.** F	**12.** F

C. Less than, Equal to, Greater than:

1. EQ	**2.** GT	**3.** GT	**4.** EQ	**5.** MI

D. Problems:

1. $Cr_2O_7^{2-}$ (aq) $+$ 14 H^+ (aq) $+$ 6 e^- $\rightarrow$ 2 Cr^{3+} (aq) $+$ 7 H_2O $E_{red}^\circ = 1.33\,V$

2. $2\,Br^-$ (aq) $\rightarrow$ Br_2 (ℓ) $+$ 2 e^- $E_{ox}^\circ = -1.077\,V$

3. $Cr_2O_7^{2-}$ (aq) $+$ 14 H^+ (aq) $+$ 6 Br^- (aq) $\rightarrow$ 2 Cr^{3+} (aq) $+$ 3 Br_2 (ℓ) $+$ 7 H_2O $E^\circ = 0.25\,V$

4. 0.01 V

5. 1.09

6. $-1.4 \times 10^2\,kJ$

7. 2.2×10^{25}

8. 15.1 g Ag

9. 0.057 L

19

Nuclear Chemistry

I. Radioactive Nuclei
 A. Mode of decay
  1. Alpha emission
   a. An α–particle has a charge of _____ and a mass of _____ .
   b. When an α–particle is emitted the atomic number of the element is

   _____ and its mass number is _____ .
   c. Example of α–radiation:

   _____

  2. Beta emission
   a. β–particles are _____ particles and have iden-

   tical properties to those of _____ .

   b. β–emission converts a _____ in the nucleus to a _____ .
   c. When a β–particle is emitted the atomic number of the element is

   _____ and its mass number is _____ .
   d. Example of β–emission:

   _____

  3. Gamma radiation
   a. These particles consist of _____ .

   b. Gamma emissions result from _____ .
   c. When a γ–particle is emitted the atomic number of the element is

   _____ and its mass number is _____ .
  4. Positron emission
   a. Positrons are identical to _____ except that a positron

   has a charge of _____ rather than –1.

b. Symbol for a positron: _____

c. Example of positron emission:

_____

5. K–electron capture

    a. An electron in energy level n = 1 "falls" into the nucleus.

    b. The result of K–electron capture is the same as _____ .

    Mass number _____ while atomic number _____ by one.

    c. Example of K–capture:

_____

6. *Exercises*

    a. Write a nuclear reaction to represent α–particle emission by $^{222}_{86}$Rn.
    Solution: Since we know that α–particles are identical to the nuclei of helium atoms, we can write

$$^{222}_{86}\text{Rn} \rightarrow {}^{4}_{2}\text{He}$$

Remember, the mass number is on top and the atomic number is at the bottom. Now, you have to balance the atomic number as well as the mass number on both sides of the equation. We have a mass number of 222 on the left and 4 on the right. So we need an element with a mass number of 222 - 4 = 218. Doing the same for the atomic number, we get

$$^{222}_{86}\text{Rn} \rightarrow {}^{4}_{2}\text{He} + {}^{218}_{84}$$

To determine the identity of the element, we look at the periodic table to see what element has the atomic number of 84. That turns out to be polonium (Po). Thus, our final equation is

$$^{222}_{86}\text{Rn} \rightarrow {}^{4}_{2}\text{He} + {}^{218}_{84}\text{Po}$$

    b. Write a balanced nuclear reaction to represent the decay of bismuth–215 to polonium–215. **(E1)**

B. Bombardment reactions
   1. Bombardment reactions are those in which _____ .
   2. Bombarding particles
      a. Neutron
         (1) atomic symbol for a neutron: _____
         (2) Example of a neutron bombardment reaction:

         _____

      b. A charged particle
         (1) Charged particles are accelerated to high velocities in _____

             or _____ fields.
         (2) When a charged particle has a high velocity, it acquires enough energy

             to bring about a nuclear reaction despite _____ with
             components of the atom.
         (3) Example of a bombardment with a charged particle:

         _____

   3. Exercises
      a. Write the reaction that represents the bombardment of Ca–41 with a β–particle.
         Solution: We start by writing the nuclear symbols of the reactants. Ca–41 has nuclear symbol $^{41}_{20}Ca$, while that of a β–particle is $^{0}_{-1}e$, so we have for the reactants:

         $$^{41}_{20}Ca + ^{0}_{-1}e \rightarrow$$

         We now determine the nuclear symbol of the product. It must have a mass number of 41 (41 + 0), and an atomic number of 19 (20 − 1). The element with atomic number 19 is K. The reaction therefore is

         $$^{41}_{20}Ca + ^{0}_{-1}e \rightarrow ^{41}_{19}K$$

      b. Write the nuclear equation for the bombardment of curium–246 (Cm) with carbon–13 to produce nobelium–254 (No). **(E2)**

C. Applications
   1. Medicine

   Example : _____

   2. Chemistry
   Neutron activation analysis
   Brief description:

   3. Commercial applications
      a. smoke alarm
      b. food irradiation

II. Kinetics of Radioactive Decay
   A. Instruments for measuring decay
      1. Geiger counter
      2. Scintillation counter
   B. First-order rate reaction
      1. Rate vs. concentration (amount present)
         a. Mathematical expression

$$\text{rate} = k\,N_t$$

where k = _____ with units _____

N_t = _____ with units _____

   b. In radioactive decay, the rate is often referred to as _____ and
      it is most often expressed in the number of atoms decaying in unit time.
   c. Activity is often given in curies.

   1 curie (Ci) = _____

   d. *Exercises*
      (1) Sulfur-35 is a radioactive isotope. A 0.58-mg sample has an activity
         of 46.2 Ci. What is the rate constant k for its decay?
         Solution: Since we are asked for k, we must be given both rate and
         N_t. The rate is expressed in its activity with curies as unit. We thus
         convert curies to atoms.

$$46.2\,\text{Ci} \times \frac{3.700 \times 10^{10}\ \frac{\text{atom}}{\text{s}}}{1\ \text{Ci}} = 1.71 \times 10^{12}\ \frac{\text{atom}}{\text{s}}$$

N_t is given in terms of milligrams. We convert that to atoms.

$$0.58 \times 10^{-3}\,\text{g} \times \frac{1\,\text{mol}}{32.06\,\text{g}} \times \frac{6.022 \times 10^{23}\ \text{atoms}}{1\,\text{mol}} = 1.08 \times 10^{19}\ \text{atoms}$$

We now substitute into the formula

$$rate = k\,N_t$$

and get

$$k = \frac{1.71 \times 10^{12}\ \dfrac{atom}{s}}{1.08 \times 10^{19}\ atoms} = 1.58 \times 10^{-7}\,s^{-1}$$

(2) Sodium–24 has a decay constant of $1.23 \times 10^{-5}\,s^{-1}$. What is the activity, in curies, of 0.500 mg of Na–24? **(E3)**

2. Concentration vs. time
   a. Mathematical expression

$$\ln \frac{X_o}{X} = k\,t$$

where X_o is _____

X is _____

k is _____

t is _____

   b. Half-life
   Relation of half-life to decay constant

$$k = \frac{0.693}{t_{\frac{1}{2}}}$$

c. *Exercises*
   (1) The half-life of radioactive sulfur–35 is 87.9 days.
      (a) What is the rate constant?
         Solution: For this, we simply use the equation relating the rate constant and half-life.

$$k = \frac{0.693}{87.9\ days} = 7.88 \times 10^{-3}\,day^{-1}$$

   (b) After what time will only one fourth of the sample remain radioactive?

Solution: For this part of the problem, we use the equation

$$\ln \frac{X_o}{X} = kt$$

where X_o is the amount of radioactive material initially present and X is the amount of material remaining after time t. In this case X is 1/4 of X_o; that is $X = X_o/4$. We use the rate constant k calculated above: 7.88×10^{-3} day^{-1}. Our unknown variable is t. Substituting into the rate equation, we get

$$\ln \frac{X_o}{\dfrac{X_o}{4}} = 7.88 \times 10^{-3} \text{ day}^{-1} \times t$$

$$\ln 4 = 7.88 \times 10^{-3} \text{ day}^{-1} \times t$$

$$t = \frac{1.386}{7.88 \times 10^{-3} \text{ day}^{-1}} = 176 \text{ days}$$

(2) The rate constant for the radioactive decay of cobalt–60 is 0.13/yr.
(a) Calculate the half-life of cobalt–60.

(b) What percent of the sample will remain after ten years? **(E4)**

C. Applications
   1. Age of rocks
   Measure the ratio of U–238 to Pb–206 to calculate the time elapsed since the rock solidified.
   Equation for U–238 to Pb–206 conversion:

   _____

   2. Age of organic material
   a. Carbon dating
   Compares the ratio of C–14 to C–12 of the organic material whose age has to be determined with the C–14/C–12 ratio of living plants.
   b. *Exercises*
   (1) An Indian artifact is estimated to be 3.50×10^3 years old. Calculate the relation between C–14/C–12 of the artifact and that of a tree living today. The half-life of C–14 is 5720 years.
   Solution: First, we calculate the rate constant of the reaction using the equation

$$k = \frac{0.693}{t_{\frac{1}{2}}}$$

Substituting 5720 for the half life ($t_{\frac{1}{2}}$), we get

$$k = \frac{0.693}{5720 \text{ yr}} = 1.21 \times 10^{-4} \text{ yr}^{-1}$$

In order to compare the ratio of C–14 to C–12 for the artifact and the living tree, we first calculate X_o/X. Substituting into the equation

$$\ln \frac{X_o}{X} = kt$$

we have

$$\ln \frac{X_o}{X} = (1.21 \times 10^{-4} \text{ yr}^{-1})(3.50 \times 10^3 \text{ yr}) = 0.424$$

Now, taking inv ln of both sides, we obtain

$$\frac{X_o}{X} = 1.53 \qquad X = 0.654\, X_o$$

Hence the C–14 to C–12 ratio in the artifact is 0.654 times the ratio in a living tree.

(2) Calculate the age of a piece of papyrus claimed to have been found in an Egyptian pyramid. Carbon dating shows a C–14 to C–12 ratio that is 0.695 times that in a plant living today. Is the papyrus authentic? (The pyramids were built between 2600 B.C. and 1700 B.C.) **(E5)**

III. Mass-Energy Relations
A. Quantitative relationship
1. Equation

$$\Delta E = \Delta m c^2$$

To calculate Δm, use the nuclear masses given in Table 19.2 of your text. Do not use the atomic masses given in the periodic table. Those masses are average atomic masses.

$$\Delta E = \text{energy of products} - \text{energy of reactants}$$
$$\Delta m = \text{mass of products} - \text{mass of reactants}$$

2. Units

In dealing with mass-energy relations, we most often express energy in kilojoules and mass change in grams. The equation then is

$$\Delta E \text{ (in kJ)} = (9.00 \times 10^{10}) \Delta m \text{ (in grams)}$$

3. Spontaneity of a nuclear reaction

The reaction is spontaneous if $\Delta m < 0$ or $\Delta E < 0$.

4. *Exercises*

    a. Calculate the energy (in kJ) released when one mole of $^{10}_{5}B$ reacts as follows:

$$^{10}_{5}B + \,^{4}_{2}He \;\rightarrow\; \,^{13}_{6}C + \,^{1}_{1}H$$

Solution: To solve this problem, we have to look up the nuclear masses in Table 19.3. You will notice that there are, for example, two entries for B. Choose the one with atomic number 5 (the subscript in $^{10}_{5}B$) and atomic mass 10 (the superscript in $^{10}_{5}B$). Be careful in looking up atomic masses. Do not round off as the mass changes are usually very small that involve the fourth or fifth decimal place. The atomic masses are

$$^{10}_{5}B = 10.01019; \quad ^{4}_{2}He = 4.00150; \quad ^{13}_{6}C = 13.00006; \quad ^{1}_{1}H = 1.00728$$

Substituting into the equation

$$\Delta m = \left(\sum m_{products}\right) - \left(\sum m_{reactants}\right)$$

we get

$$\Delta m = (13.00006 + 1.00728) - (10.01019 + 4.00150) = -0.00435 \text{ g}$$

Substituting now into the equation

$$\Delta E = (9.00 \times 10^{10}) \Delta m$$

we obtain

$$\Delta E = (9.00 \times 10^{10})(-0.00435) = -3.92 \times 10^8 \text{ kJ}$$

This means that -3.92×10^8 kJ are evolved into the surroundings when *one mole* of boron–10 reacts with α–particles.

b. Lithium-6 $\left(^6_3\text{Li}\right)$ is bombarded with neutrons to obtain α–particles and tritium $\left(^3_1\text{H}\right)$.

(1) Write the equation for this nuclear reaction.

(2) Calculate the amount of energy evolved when 10.0 g of $^6_3$Li is bombarded by neutrons. **(E6)**

B. Nuclear binding energy
1. A nucleus weighs _____ the individual protons and neutrons of which it is composed.
2. The mass of nucleons (protons and neutrons) minus the mass of the nucleus listed in Table 19.3 gives a quantity referred to as the mass defect.

mass defect $= \Delta m =$ mass of nucleons $-$ mass of nucleus

3. The corresponding energy is known as the binding energy.

$$\text{binding energy} = (\text{mass defect}) \times 9.00 \times 10^{10} \frac{\text{kJ}}{\text{g}}$$

4. The binding energy is the energy required to decompose one mole of nucleus into nucleons (protons and neutrons).
5. The binding energy of a nucleus is a measure of its stability. The greater the binding energy, the more difficult it is to decompose the nucleus into its component nucleons.

IV. **Nuclear Fission**
   A. Fission process
      Give a brief description:

      _____

      _____

      _____

   B. Fission energy
      Used in nuclear reactors
      Give a brief description:

      _____

      _____

      _____

V. **Nuclear Fusion**
   A. Definition

      _____

   B. Why is it attractive now to people?

      _____

      _____

      _____

SELF-TEST
A. Multiple Choice:

1. When $^{232}_{90}\text{Th}$ undergoes alpha emission, the product nucleus is

   a. $^{232}_{88}\text{Ra}$ **b.** $^{228}_{88}\text{Ra}$ **c.** $^{228}_{90}\text{Th}$ **d.** $^{232}_{91}\text{Pa}$ **e.** $^{232}_{89}\text{Ac}$

2. The number of balanced equations from the set below is

   $$^{1}_{0}n + ^{235}_{92}U \rightarrow ^{90}_{36}Rb + ^{144}_{58}Cs + 2\,^{1}_{0}n$$

   $$^{1}_{0}n + ^{238}_{92}U \rightarrow ^{239}_{94}Pu + 2\,^{0}_{-1}e$$

   $$^{2}_{1}H + ^{2}_{1}H \rightarrow ^{4}_{2}He$$

   $$^{226}_{88}Ra \rightarrow ^{227}_{86}Rn + ^{4}_{2}He$$

   a. 0   **b.** 1   **c.** 2   **d.** 3   **e.** 4

3. The half-life of an element is 20 years. What fraction of the element has decayed after 40 years?

   a. 1/4   **b.** 1/2   **c.** 3/4   **d.** 7/8   **e.** can't tell

4. The correct statement in the following list is:
   a. Fission involves electron absorption.
   b. Fission and fusion liberate energy.
   c. Fusion liberates energy, fission absorbs energy.
   d. Fission liberates energy, fusion absorbs energy.
   e. None of the above statements is true.

5. In a nuclear reaction, when the atomic number is lowered by one and the mass number is unchanged, the particle emitted is
   a. a positron   **b.** an α–particle   **c.** a β–particle   **d.** a neutron

6. Which of the following statements is NOT true about the fission of uranium–235?
   a. The reaction is initiated by the capture of a proton.
   b. On the average, at least 2 neutrons result from the fission of every uranium nucleus.
   c. The fission products emit beta and gamma radiation.
   d. The sum of the masses of the products is less than the sum of the masses of the reactants.

7. Biological damage by nuclear radiation is caused by its ability to

   (1) produce holes in biological membranes.
   (2) cause mutations.
   (3) ionize organic molecules.
   (4) burn cells.
   (5) initiate a nuclear explosion.

   The true statements are
   a. (1),(2)   **b.** (2),(3)   **c.** (1),(2),(3)   **d.** (3),(4)   **e.** (3),(4),(5)

8. How many of the following statements are true?

   — Gamma radiation consists of photons of light.
   — Emission of a proton increases both the atomic number and mass number by one.
   — In nuclear fusion, two light nuclei come together.
   — A nuclear reaction involves rearranging the electrons in the energy level farthest from the nucleus.

   a. 0   **b.** 1   **c.** 2   **d.** 3   **e.** 4

9. In a certain nuclear reaction the mass of products is slightly greater than the mass of reactants. This means that
   a. the reaction is spontaneous.
   b. the reaction is exothermic.
   c. the reaction is endothermic.
   d. there is experimental error. The masses of products and reactants must be the same.

10. If U–238 has a half-life of 4.5×10^9 yr and Rb–87 has a half-life of 5.7×10^{10} yr, how do their rate constants compare?
   a. Rate constant of U–238 < rate constant of Rb–87.
   b. Rate constant of U–238 > rate constant of Rb–87.
   c. Rate constant of U–238 = rate constant of Rb–87.
   d. Cannot tell.

B. Problems:

   Consider U–238, which undergoes natural radioactive decay to form thorium–234. U–238 has a half-life of 1.62×10^5 yr.
   1. Write a balanced nuclear equation for this process.

   2. What is the activity, in curies, of 1.00 mg of U–238?

3. How many grams of U–238 will remain from a one-gram sample after 2.00×10^3 yr of radioactive decay?

4. What is the energy in kilojoules associated with the decay of 1.00 mg of U–238? The nuclear mass of thorium–234 is 234.0437 g.

5. What is the mass defect of U–238?

6. What is its binding energy?

ANSWERS
Exercises:
(E1) $^{215}_{83}Bi \rightarrow {}^{215}_{84}Po + {}^{0}_{-1}e$

(E2) $^{246}_{96}Cm + {}^{13}_{6}C \rightarrow {}^{254}_{102}No + 5\,{}^{1}_{0}n$

(E3) 4.35×10^3 Ci

(E4) (a) 5.3 yr   **(b)** 27 %

(E5) 3.00×10^3 yr;   No.   3000 − 1996 = 1004. The papyrus was a living plant around 1008 B.C., long after the last pyramid was built.

(E6) (a) $^{6}_{3}Li + {}^{1}_{0}n \rightarrow {}^{4}_{2}He + {}^{3}_{1}H$ **(b)** 7.71×10^8 kJ are evolved

Self-Test
A. Multiple Choice:
1. b   **2.** c   **3.** c   **4.** b   **5.** a

6. a   **7.** b   **8.** c   **9.** c   **10.** b

B. Problems:
1. $^{238}_{92}U \rightarrow {}^{234}_{90}Th + {}^{4}_{2}He$ **2.** 9.27×10^{-6} Ci   **3.** 0.991 g

4. 1.70×10^4 kJ   **5.** 1.9353 g   **6.** 1.74×10^{11} kJ/mol

20

Chemistry of the Metals

This chapter is largely descriptive, with exercises to review concepts presented in many of the earlier quantitative chapters of this text. Since those chapters have been extensively covered in this workbook, there is no self-test. Filling in the outline from memory should be a sufficient review of the chapter.

I. **Metallurgy**
   A. Chloride ores: Na from NaCl
      1. Sodium metal is obtained from the chloride ore by _____ .
      2. Reaction at the cathode:

      3. Reaction at the anode:

      4. Overall reaction:

   B. Oxide ores
      1. Al from Al_2O_3
         Electrolytic reaction:

      2. Fe from Fe_2O_3

         a. Ore: _____

b. Reduction of Fe^{3+}

    (1) This is carried out in _____ .

    (2) The "charge" admitted to the top consists of

        (a) _____

        (b) _____

        (c) _____

    (3) Reactions

        (a) Carbon to carbon monoxide

        Equation: _____

        (b) Iron(III) oxide reacts with carbon monoxide to produce iron metal and carbon dioxide.

        Equation: _____

        (c) Decomposition of calcium carbonate to calcium oxide and carbon dioxide at $800°C$

        Equation: _____

        (d) Formation of slag by reacting calcium oxide with silicon dioxide to form liquid calcium silicate ($CaSiO_3$)

        Equation: _____

    (4) The product is called _____ .

c. Converting pig iron to steel

    (1) Remove Si, Mn, and P by reaction with O_2 to form silicon dioxide, manganese dioxide, and P_4O_{10} (s).

    (2) Lower carbon content below 2% by burning the carbon with oxygen to form CO_2.

d. Most of the steel produced today is made by a process called

_____ .

C. Sulfide Ores

  1. General process

    a. After preliminary treatment, these ores are roasted, which means

_____ .

Reaction for the roasting of ZnS:

_____

    b. The oxide produced is then reduced to _____ .

    c. If the metal ore is unstable at the roasting temperature, the products are

_____ and _____ .

Reaction for the roasting of HgS:

_____

 2. Copper from Cu_2S
     a. Principal ore: _____
     b. Steps in copper extraction
        (1) Cu_2S is concentrated by a process called _____ .
        (2) The concentrated ore is converted to metal by

 _____ .

 Reaction: _____

        (3) The solid product from step (2) is called _____ .

        (4) Copper is purified by _____ .

 anode: _____

 cathode: _____

 electrolyte: _____
 oxidation half-reaction:

 _____

 reduction half-reaction:

 _____

 overall reaction:

 _____

D. Native metals: Au
 1. The equation for the reaction that puts gold into solution as a complex ion:

 _____

 2. The equation for the reaction that converts the complex ion back to the metal:

 _____

II. Reactions of the Alkali and Alkaline Earth Metals

Learn and remember the general equations in Table 20.1 of your text!

A. Reactions with hydrogen:
 1. The metals of Group 1 form the hydride MH.
 Write the equation for the reaction between sodium and hydrogen:

 _____

 2. The metals of Group 2 form the hydride MH_2.
 Write the equation for the reaction between barium and hydrogen:

 _____

 3. These hydrides when added to water react to give metal ion, the hydroxide ion and hydrogen gas.
 Write the equation for the reaction between potassium hydride and water:

 _____

B. Reactions with water

    1. All Group 1 metals react with water to form the hydroxide ion, the metal ion, and hydrogen gas.
Write the equation for the reaction between lithium and water:

_____

    2. Only the Group 2 metals Ca, Sr, and Ba react with water to form the same products as the Group 1 metals.
Write the equation for the reaction between calcium and water:

_____

C. Reaction with oxygen

    1. Formation of oxides:

       a. Group 1: Only lithium forms Li_2O by direct reaction with oxygen.

       b. All Group 2 metals form the oxide MO when reacted with oxygen.

    2. Formation of peroxides
The anion is O_2^{2-}. The most important peroxides are K_2O_2 and Na_2O_2. A peroxide reacts with water to give the metal ion, the hydroxide ion, and hydrogen peroxide (H_2O_2).

    3. Formation of superoxides
The anion is O_2^-.
What is the use for KO_2? _____

III. Redox Chemistry of the Transition Metals

    A. Reactions of the transition metals with oxygen

    1. Table 20.2 lists the formulas of the oxides formed when the more common transition metals react with oxygen.

    2. Transition metals are usually present as ions with a _____ and/or _____ charge.
An exception is _____ .

    3. Reaction for the preparation of silver(I) oxide by treatment of the silver salt with a strong base:

_____

    4. Reaction for the preparation of cobalt(II) oxide by heating $CoCO_3$:

_____

    5. What is a nonstoichiometric oxide?

_____

    B. Reaction of transition metals with acids

    1. Any metal with a _____ can be oxidized by the H^+ ions in a 1 M solution of a strong acid.

    2. Copper has a negative E_{ox}° and therefore cannot be oxidized by the H^+ ion of a strong acid. Nitric acid however, can oxidize Cu. The reaction is

_____ .

3. Gold is not oxidized by HNO_3, but it can be brought into solution by *aqua*

regia which is a mixture of _____ .
Reaction between gold and aqua regia:

_____

C. Equilibria between different cations of a transition metal
   1. Examples of a transition metal with more than one cation:

   2. Cations for which E°_{red} is a large positive number are _____ and

   unstable in _____ .
D. Oxoanions of the transition metals
   1. CrO_4^{2-}
      a. Oxidation number of chromium in the anion: _____

      b. Color: _____

      c. Reaction in acid solution:

      _____

      d. The CrO_4^{2-} is stable in _____ solution.
   2. $Cr_2O_7^{2-}$
      a. Oxidation number of chromium in the anion: _____

      b. Color: _____

      c. Reaction in acid solution:

      _____

      d. Reaction when $(NH_4)_2Cr_2O_7$ is ignited:

      _____

   3. MnO_4^{-}
      a. Color: _____

      b. Uses: _____

      c. Reaction in acid solution:

      _____

      d. Reaction in basic solution:

      _____

IV. Alloys
   A. Definition:

   _____

   B. Types of alloys
      1. Solid solutions
         a. The two metals are completely soluble in the solid as well as the liquid state.
         b. Solid solutions are rare because complete solubility in the solid requires that both metals have

            (1) the same type of _____ ;

            (2) similar _____ .
         c. Example of a solid solution:

            _____

         d. Interstitial solid solution

            (1) Definition: _____

            (2) Example: _____
      2. Heterogeneous mixture
         Description:

   C. Alloys can contain intermetallic compounds.
      1. What are intermetallic compounds?

         _____

         _____

      2. Effects of alloying

         a. lowers _____

         b. increases _____

         c. lowers _____

21

Chemistry of the Nonmetals

Chapter 21, like Chapter 20, is descriptive. Thus, there is no self-test at the end of the chapter. Review for this chapter by filling out the outline from memory.

I. **The Elements and Their Preparation**
   A. Properties of eight common nonmetals
      Study Table 21.1 of your text without memorizing it.
   B. Chemical reactivity
      1. Nitrogen
         a. Its inertness is due to _____
            _____ .
         b. Equation for the detonation of sodium azide:

         c. Use for sodium azide:

            _____
      2. Fluorine
         a. Most reactive of elements because _____
            _____
         b. It combines with every element in the periodic table except _____ .
         c. F_2 (g) is ordinarily stored in _____ .
      3. Chlorine
         a. Reacts with nearly all metals although _____ is often required.
         b. Chlorine disproportionates in water forming _____
            and _____ .

Reaction:

_____

C. Occurrences and preparations
1. Nitrogen and oxygen
   a. Obtained from liquid air
   b. Brief description of the process

2. Halogens
   a. Fluorine
      (1) They are found in the mineral _____ .
      (2) They are obtained by electrolytic oxidation.

         (a) Developed the process: _____

         (b) Electrolyte used: _____
         (c) Function of each component of the electrolyte mixture:

             _____

             _____

   b. Chlorine
      (1) It is found in _____ .
      (2) It is obtained by electrolysis
   c. Bromine and iodine
      (1) It is found in _____ .
      (2) It is prepared by chemical oxidation.
         Reactions:

         _____

         _____

      (3) Common oxidizing agent used: _____
II. **Hydrogen Compounds of Nonmetals**
   A. Ammonia – NH_3
      1. Reaction used in the Haber process:

      2. Most common use is in the manufacture of fertilizers.
      3. Equation for a reaction showing how it acts as a Bronsted-Lowry base:

4. Equation for the formation of hydrazine:

B. Hydrogen sulfide – H_2S
    1. Bronsted-Lowry acid
       Reaction with water:

_____

    2. Precipitation reaction with a metal:

_____

    3. Reaction with oxygen:

_____

C. Hydrogen peroxide – H_2O_2
    1. Reaction for the oxidation of H_2O_2 to H_2O :

_____

    2. Reaction for the reduction of H_2O_2 to O_2 :

_____

    3. Reaction for the disproportionation of H_2O_2 to H_2O and O_2 :

_____

D. Hydrogen fluoride and hydrogen chloride
    1. Compare the acid strengths of aqueous solutions of HCl and HF.

    2. Reaction with OH^-
       a. HCl (aq)

_____

       b. HF (aq)

_____

    3. Reaction with CO_3^{2-}
       a. HCl (aq)

_____

       b. HF (aq)

_____

    4. Uses of HF:

    5. Storage of HF: _____

III. Oxygen Compounds of Nonmetals
    A. Molecular structure
        1. Oxides of nitrogen that are paramagnetic are _____ ,
             _____ .

        2. Equation for the formation of N_2O_3 :

        _____

        Lewis structure of N_2O_3 :

        3. N_2O
            a. Common name: _____
            b. Uses: _____
            c. Resonance forms:

    B. Reactions of nonmetal oxides with water
        1. Acid anhydride

        Definition: _____

        Example: _____
        2. Formation of H_2SO_4 from its anhydride:

        _____

        3. Formation of HNO_3 from its anhydride:

        _____

        4. Formation of H_3PO_4 from its anhydride:

        _____

IV. Oxoacids and Oxoanions
    A. Acid Strength
        1. The value of the dissociation constant K_a for an oxoacid increases with

            a. _____

            b. _____

        2. The electron density around the oxygen atom is decreased when

            a. _____

            b. _____

B. Oxidizing and reducing strength

1. A species in which a nonmetal is in its highest oxidation state can act only as

   _____ and never as _____ .

2. A species in which a nonmetal is in an intermediate highest oxidation state

   can act either as _____ or _____ .

3. When oxidation and reduction occur together, _____

   results. This happens when E°_{ox} and E°_{red} are _____ 0.

4. The oxidizing strength of an oxoacid or oxoanion is greatest at _____

   $[H^+]$ or _____ pH.

C. Nitric Acid

1. Reaction of dilute (6 M) HNO_3 with $Al(OH)_3$ (s):

   _____

2. Reaction with $CaCO_3$:

   _____

3. Reaction of concentrated (16 M) HNO_3 with Cu (s)

   _____

4. Reaction of dilute (6 M) HNO_3 with Cu (s):

   _____

5. Reaction of dilute (6 M) HNO_3 with Zn (s):

   _____

6. Reaction to show why colorless concentrated HNO_3 turns yellow:

   _____

7. HNO_3 reacts with proteins, like skin, to give a yellow material called

   _____ .

D. Sulfuric Acid

1. Two step ionization of H_2SO_4:

   _____

   _____

2. Reaction for the reduction of H_2SO_4 to SO_2 (g):

   _____

3. Reaction of hot concentrated H_2SO_4 with Cu:

   _____

4. Reaction of hot dilute H_2SO_4 with Zn:

   _____

5. Reaction to show the dehydrating effect of H_2SO_4 on table sugar, $C_{12}H_{22}O_{11}$:

   _____

E. Phosphoric Acid

    1. Two step ionization of H_3PO_4:

_____

_____

_____

    2. The dominant species at pH < 2 is _____ ; at pH 2 – 7, it is _____ ; at pH 7 – 11, it is _____ ; at pH > 11, it is _____ .

22

Organic Chemistry

Chapter 22 is a descriptive chapter. Like the earlier descriptive chapters, this one has no self-test or quantitative problems. Review for this chapter by filling in the blanks in the outline from memory.

I. **Common Features of Organic Compounds**

    A. Organic compounds are _____ rather than ionic.

    B. Each carbon atom has a total of _____ covalent bonds.

    C. Carbon atoms may be bonded to each other or to other _____ atoms. In all these compounds:

        1. A hydrogen or halogen atom forms _____ covalent bond(s).

        2. An oxygen atom forms _____ covalent bond(s).

        3. A nitrogen atom forms _____ covalent bond(s).

II. **Saturated Hydrocarbons – Alkanes**

    A. Definition

    In saturated hydrocarbons, also known as _____ , the carbon-

    carbon bonds are _____ bonds.

    B. Structural isomers

    Definition: _____

    _____

    Example: _____

C. Nomenclature

    1. Straight chain

        a. Made up of only one word

        b. Name of alkane and alkyl group with

           1 C _____

           2 C _____

           3 C _____

           4 C _____

           5 C _____

           6 C _____

           7 C _____

           8 C _____

    2. Branched chain

        a. Suffix identifies _____

        To find the suffix, count the number of atoms in the _____ .

        b. Prefix identifies _____ and indicates by number the

        carbon atom where _____ occurs.

        c. If the same alkyl group is at 2 branches, then the prefix _____
        is used.

        d. The number in the name is made as small as possible. To do that, one
        can start either at the right end or at the left end.

D. Major sources of alkanes

    1. natural gas

    2. petroleum

III. Unsaturated Hydrocarbons

  A. Alkenes

    1. Definition: _____

    2. Simplest alkene

      Formula: _____ ; name: _____

      structure: _____

    3. Preparation of ethene

      Equation: _____

B. Alkynes

   1. Definition: _____

   2. Acetylene

   a. Formula: _____

   b. Bond angle: _____

   c. Number of sigma bonds: _____ ; number of pi bonds: _____

   d. Hybridization: _____

   e. Show that it is thermochemically unstable with respect to decomposition
         to the elements.

   _____

   f. Thermochemical equation for the combustion of acetylene

   _____

IV. Aromatic Hydrocarbons

   A. Hydrocarbons of this type are referred to as _____, and can

   be considered to be derivatives of _____

   B. Derivatives of benzene

   When 2 groups are attached to the benzene ring, _____ isomers

   are possible. They are designated by the prefixes _____ , _____ ,

   and _____ . Numbers can also be used.

   C. Condensed ring structures

   The simplest compound of this type is _____ , with formula

   _____ .

V. Functional Groups

   A. Alcohols

   1. An alcohol can be considered to be derived from a hydrocarbon by replacing

   one or more of the H atoms by _____ .

   2. Naming

   The suffix _____ is substituted for the _____ suffix of the
   corresponding alkane.

   3. Methanol

   a. Common name _____

   formula _____

b. Preparation
  (1) It is produced from _____ , a mixture of

  _____ and _____ .

  Equation: _____

  (2) It is also formed as a by-product when charcoal is made by heating

  _____ in the absence of _____ .

4. Ethanol
  a. Common name: _____

  Formula: _____

  b. Preparation

  Fermentation of _____ or _____

  c. Denatured alcohol is _____

  _____ .

  d. Preparation of industrial alcohol

  Reaction: _____

  e. The proof of an alcoholic beverage is _____ .

5. Example of an alcohol with 2 OH groups in one molecule:

  _____

B. Ethers
  1. General structure for ethers: _____

  2. How are ethers named? _____

  3. Compare the boiling point and the water solubility of ethers, alcohols and alkanes of similar molar mass.

  _____

  4. Do ethers show hydrogen bonding in water? _____

  5. Write the formula of the ether once used as an anesthetic.

  _____

C. Aldehydes and Ketones
  1. Both compounds contain _____ group.

  2. How do aldehydes differ from ketones?

  _____

  3. Simplest aldehyde: formaldehyde

  a. Formula: _____

  b. IUPAC name: _____

4. Simplest ketone: acetone

    a. Formula: _____

    b. IUPAC name: _____

    c. Bond angle around central carbon atom: _____

D. Carboxylic acids

    1. A carboxylic acid can be considered to be derived from hydrocarbons by re-

    placing one or more H atoms by a _____ group,

    often abbreviated _____ .

    2. Naming

      a. The suffix _____ is added to the stem of the corresponding
      alkane.

      b. The common name for methanoic acid is _____ . Its

      formula is _____ .

      c. The common name for ethanoic acid is _____ . Its

      formula is _____ . It is the active ingredient in _____

      and is responsible for its _____ taste.

    3. Carboxylic acids act as weak acids in water solution.

      a. Dissociation equation for RCOOH (aq):

      _____

      b. The more _____ the R group, the stronger the acid.
      c. _____ is one of the strongest organic acids.

    4. Treatment of a carboxylic acid with the strong base NaOH forms the

    _____ of the acid.

    Reaction of acetic acid with NaOH:

    _____

    5. An example of a long-chain carboxylic acid is _____ .
    Its sodium salt is a soap. The long hydrocarbon group is a good solvent for

    _____ , while the ionic COO^- group gives _____ .

E. Esters

    1. An ester is formed by the reaction between _____ and

    _____ .

    2. Its functional group is _____ .

    3. Reaction between acetic acid and methanol:

    _____

F. Amines
   1. General formula: _____
   2. Reaction of an amine (RNH_2) with a strong acid:

   _____

   3. Reaction of methylamine with a weah acid, e.g., acetic acid:

   _____

VI. Isomerism in Organic Compounds
   A. Structural Isomerism
   This type of isomerism is discussed in Section 22.1 of your text.
   B. Geometric isomerism
      1. How does geometric isomerism result?

      _____

      2. Two types of geometric isomers
         a. *trans* form
            Definition: _____

         b. *cis* form
            Definition: _____

   C. Optical isomerism
      1. Definition: _____

      _____

      2. When a molecule exists in two different forms that are not superimposable
      mirror images, it is said to be _____ .
      3. The two different forms are referred to as _____ .
      4. The carbon atom to which the four different groups are bonded is called

      _____ .

VII. Organic Reactions
   A. Addition Reactions
      1. In an addition reaction, a small molecule adds across a _____

      or a _____ bond.

      2. Liquid fats are converted to solid fats by _____

      where treatment with _____ under a suitable catalyst

      converts _____ to _____ .
      3. The process of adding water to an alkene to produce an alcohol is called

      _____

Organic Chemistry   **415**

B. Elimination reactions
An elimination reaction involves removing two groups from _____

thus converting a saturated molecule to _____ .
C. Condensation reactions
1. A condensation reaction occurs when _____ combine

by _____ a small molecule. The molecules that combine
may be the same or different.

2. Condensing 2 alcohols produces _____.

3. Condensing an alcohol and a carboxylic acid produces _____.
D. Substitution reactions

1. A substitution reaction is one in which _____

2. Chlorination of alkanes
Equation for the reaction between ethane and chlorine:

_____

3. Nitration of aromatic hydrocarbons
Equation for the nitration of benzene:

_____

4. Formation of an alcohol from an alkyl halide
Equation for the reaction between ethylbromide and a strong base:

_____

23

Organic Polymers, Natural and Synthetic

Chapter 23 is a descriptive chapter. Like the earlier descriptive chapters, this one has no self-test or quantitative problems. Review for this chapter by filling in the blanks in the outline from memory.

I. Synthetic Addition Polymers

   A. Nature of addition polymers

   1. Definition: _____

   _____

   2. They are derived from monomers containing a _____

   which is converted to _____ upon polymerization.

   B. Polyethylene
      1. Monomer: _____
      2. Reaction representing the polymerization process:

      _____

      3. Branched polyethylene has neighboring chains arranged _____ .

      This produces a _____ solid.
      4. Linear polyethylene has almost no branched chains. Neighboring chains line up nearly _____ to each other.

C. Polyvinyl chloride

    1. Monomer: _____

    2. Structure

        a. The CH_2 group is referred to as _____ .

        b. The CHCl group is referred to as _____ .

    3. Types of polymers formed

        a. Head-to-tail polymer

        There is a _____ atom on every other _____ atom in the chain.

        b. Head-to-head or tail-to-tail polymer

        Cl atoms occur in _____ on _____ carbon atoms in the chain.

        c. Random polymer

D. Teflon

    1. Formula for Teflon: _____

    2. Why is Teflon impervious to chemical attack?

    _____

    3. Other properties of Teflon

        a. _____

        b. _____

    4. Main commercial use of Teflon: _____

II. Synthetic Condensation Polymers

    A. Formation of condensation polymers

        1. The monomer split out in the condensation is most often _____ .

        2. The monomers involved must have _____

        3. Functional groups most often present are _____ , _____ , and _____ .

    B. Polyesters

        1. Monomers are

            a. _____

            b. _____

        2. Poly (ethylene terephthalate) (PET)

            a. Its monomers are _____ and _____ .

            b. The trade name of fabric with PET is _____ .

            c. Magnetically coated PET is known as _____ .

B. Polyamides

    1. Its monomers are _____ and _____ .

    2. An example of a polyamide is Nylon-66.

        a. What do the numbers 66 indicate?

        _____

        b. What causes the elasticity of nylon fibers?

        _____

III. Carbohydrates

    A. General formula: _____

    B. Monosaccharides

        1. Definition: _____

        _____

        2. Glucose

            a. Structure as a solid

            b. In water solution, it exists as _____ .

            c. _____ atoms make up 5 corners. The sixth corner has the

            _____ atom.

            d. There is _____ group bonded to carbon atom 5.

            e. There is a _____ group bonded to each carbon atom.

            f. There is an OH group bonded to carbon atoms _____ .

            g. The bonds to the H and OH groups from the ring carbon atoms are oriented in the plane of the ring, _____, or perpendicular to it,

            _____ .

            h. In β-glucose, all 4 OH groups are _____ .

            i. In α-glucose, the OH group on carbon 1 is _____ .

            All the rest are _____ .

    B. Disaccharides

        1. Definition: _____

        _____

        2. Maltose

            a. It is a decomposition product of _____ .

    b. It is a dimer made up of _____ molecules combined

    _____ . Carbon 1 of one molecule is joined through

    a(n) _____ atom to carbon _____
of the second molecule.

    c. It can be decomposed back to glucose molecules by the enzyme

    _____ or by heating with _____ .

3. Sucrose

    a. The monomers that make up sucrose are _____ .

    and _____ .

    b. Sucrose can be decomposed back to its monomers by the enzyme

    _____ .

D. Polysaccharides

  1. Starch

    a. It is found in plants primarily in _____

    and _____ .

    b. It is a mixture of two types of _____ polymers. One of

    these is _____ and consists of long single chains of

    _____ units joined _____ .

    The other polymer is _____ which has a

    _____ structure.

  2. Cellulose

    a. It contains _____ chains of glucose units.

    b. It differs from starch in _____ .

    c. Its molecular structure allows for _____ between poly-
mer chains.

IV. Proteins

  A. Functions

    1. Fibrous proteins: _____

    _____

    2. Proteins in body fluids: _____

    _____

    3. Hormones: _____

    _____

B. α-Amino acids

1. They have a _____ group attached to the carbon atom

   (α-carbon) adjacent to a _____ group.

2. Structural formula of an amino acid:

3. In all the amino acids, except _____, the α-carbon is

   chiral. This means that most of the α-amino acids have _____ .

4. Acid-base properties - zwitterion

   a. Definition: _____

   _____ .

   b. It behaves like a _____ molecule. Within it, there is +1

   charge at the _____ atom and a −1 charge at one of

   _____ atoms of the COOH group. Overall the net

   charge is zero.

   c. At low pH, the _____ grup picks up a proton and the

   product has _____ charge.

   d. At high pH, the _____ grup loses up a proton and the

   product has _____ charge.

   e. Zwitterions are capable of acting either as a Bronsted acid or base.

   f. The pH at which the maximum concentration of zwitterion occurs is

   called _____ .

   g. For an amino acid where there is only one COOH and one NH$_2$ group, the
   pH at the isoelectric point can be calculated using the formula

   _____ .

C. Polypeptides

1. The α-amino acid s undergo _____ to produce

   polypeptides. They contain the group _____ commonly

   called _____ .

2. Polypeptides are written by starting the far left of the amino acid with a free

_____ . The amino acid with a free _____

ends the polypeptide structure.

C. Protein Structure

    1. Primary structure

        a. It is the sequence in which the amino acids are arranged.

        b. It is determined by two methods

            (1) End group analysis

                Description of the process:

            _____

            _____

            (2) Hydrolysis of the protein chain into fragments

                Description of the process:

            _____

            _____

    2. Secondary structure

        a. When proteins align themselves, certain patterns are repeated. These repeating patterns establish its secondary structure.

        b. The nature of the pattern is determined in large part by

        _____ . Oxygen atoms on _____

        groups can interact with _____ atoms in nearby N–H groups.

    3. Tertiary structure

        a. It is the three dimensional conformation of the protein. An α-helix can be twisted, folded, or folded and twisted into a definite geometric pttern.

        b. Collagen has a teritary structure made up of _____ .